Contents

Preface   vii

1 Simplified Semiconductor Theory   1
A simple outline of atomic theory   1
Intrinsic semiconductors   5
Conductors and insulators   8
Extrinsic (impurity) semiconductors   9
The p-n junction   10
Exercises   13

2 Semiconductor Diodes   15
Construction   15
Semiconductor diode current-voltage characteristics   17
Diode load lines   22
The semiconductor diode as a switch   24
Types of diode and their application   26
Data sheets   34
Clipping and clamping circuits   36
Exercises   38

3 Bipolar Transistors   40
The action of a transistor   41
The common-base connection   43
The common-emitter connection   45
The common-collector connection   47
Transistor static characteristics   48
Thermal and frequency effects   54
The bipolar transistor as a switch   56
The construction of transistors   63
Data sheets   65
Exercises   68

4 Field-effect Transistors   75
The junction field-effect transistor   75
The metal-oxide silicon field-effect transistor   79
Static characteristics   82

Handling the MOSFET 87
The field-effect transistor as a switch 87
FET Switching circuits 88
Power MOSFETs 90
Data sheets 95
The relative merits of bipolar transistors and FETs 96
Exercises 97

5 Integrated Circuits 101
Integrated circuit components 101
The fabrication of a complete integrated circuit 108
Linear and digital integrated circuits 110
Exercises 113

6 Small-signal Audio-frequency Amplifiers 114
Principles of operation 115
Bias and stabilization 119
Bandwidth 123
Determination of gain using a load line 127
Equivalent circuits 136
Design of a single-stage audio-frequency amplifier 140
Multi-stage amplifiers 142
Exercises 143

7 Power Supplies 147
Rectifier circuits 148
Filter circuits 151
Voltage multiplication 153
Voltage regulators 154
Exercises 162

8 Combinational Digital Circuits 164
The binary code 164
Combinational logic 165
Digital integrated circuits 170
Combinational logic circuits 176
Boolean laws and identities 179
The Karnaugh map 181
Logic circuits: design examples 184
Exercises 189

9 Sequential Digital Circuits 191
Flip-flops 191
Counters 195
Shift registers 200
Exercises 203

Answers to Numerical Exercises 204

Index 205

Preface to fifth edition

The Business and Technician Education Council (BTEC) scheme for the education of electronic, electrical, computer, and telecommunication technicians includes an electronics unit at level II. For the fifth edition of Electronics 2 a complete revision has been carried out to ensure full coverage of the contents of the BTEC unit.

The book has been written on the assumption that the reader will have studied, or be concurrently studying, electrical principles to BTEC level II standard. A large number of worked examples are provided throughout the book to emphasize important concepts, and each chapter concludes with a number of exercises based on the contents of that chapter. Answers to the numerical exercises are to be found at the back of the book.

1994   D.C.G.

1 Simplified Semiconductor Theory

A semiconductor is defined as a material whose resistivity is much less than the resistivity of an insulator yet is much greater than the resistivity of a conductor, *and* whose resistivity decreases with increase in temperature. For example, the resistivity of copper is 10^{-8} ohm-metre, of quartz is 10^{12} ohm-metres, and for the semiconductor materials of interest in this chapter, that of silicon is 0.5 ohm-metre and that of germanium is 2300 ohm-metres at 27°C. To gain an appreciation of the operation of semiconductors and semiconductor devices, it is necessary to have some familiarity with the basic concepts of the atomic structure of matter.

A Simple Outline of Atomic Theory

All the substances which occur in Nature consist of one or more basic elements; a substance containing more than one element is known as a compound. An *element* is a substance that can neither be decomposed (broken into a number of other substances) by ordinary chemical action, nor made by a chemical union of a number of other substances. A compound consists of two or more different elements in combination and has properties different from the properties of its constituent parts. Water, for example, is a compound of oxygen and hydrogen. A *molecule* is the smallest amount of a substance that can occur by itself and still retain the characteristic properties of that substance, and may consist, for example, of two atoms of hydrogen and two atoms of oxygen for hydrogen peroxide, of one atom of oxygen and one atom of carbon for carbon monoxide, and of two atoms of oxygen and one atom of carbon for carbon dioxide. An *atom* is the smallest unit of which a chemical element is built. The atoms of any particular element all have the same average mass and this average mass differs from the average mass of the atoms of any other element.

The elements are grouped in an arrangement known as the Periodic Table of the Elements (Table 1.1) according to their chemical properties. Elements with similar properties are placed in the same vertical column. More than 100 elements are known to science today;

1

Table 1.1 The Periodic Table of the Elements

I	II	III	IV	V	VI	VII	VIII	0
Hydrogen (H) 1								Helium (He) 2
Lithium (Li) 3	Beryllium (Be) 4	Boron (B) 5	Carbon (C) 6	Nitrogen (N) 7	Oxygen (O) 8	Fluorine (F) 9		Neon (Ne) 10
Sodium (Na) 11	Magnesium (Mg) 12	Aluminium (Al) 13	Silicon (Si) 14	Phosphorus (P) 15	Sulphur (S) 16	Chlorine (Cl) 17		Argon (A) 18
Potassium (K) 19	Calcium (Ca) 20	Scandium (Sc) 21	Titanium (Ti) 22	Vanadium (V) 23	Chromium (Cr) 24	Manganese (Mm) 25	Iron (Fe) 26 Cobalt (Co) 27 Nickel (Ni) 28	
Copper (Cu) 29	Zinc (Zn) 30	Gallium (Ga) 31	Germanium (Ge) 32	Arsenic (As) 33	Selenium (Se) 34	Bromine (Br) 35		Krypton (Kr) 36
Rubidium (Rb) 37	Strontium (Sr) 38	Yttrium (Y) 39	Zirconium (Zr) 40	Niobium (Nb) 41	Molybdenum (Mo) 42	Technetium (Tc) 43	Ruthenium (Ru) 44 Rhodium (Rh) 45 Palladium (Pd) 46	
Silver (Ag) 47	Cadmium (Cd) 48	Indium (In) 49	Tin (Sn) 50	Antimony (Sb) 51	Tellurium (Te) 52	Iodine (I) 53		Xenon (Xe) 54
Caesium (Cs) 55	Barium (Ba) 56	Rare Earths 57–71	Hafnium (Hf) 72	Tantalum (Ta) 73	Tungsten (W) 74	Rhenium (Re) 75	Osmium (Os) 76 Iridium (Ir) 77 Platinum (Pt) 78	
Gold (Au) 79	Mercury (Hg) 80	Thallium (Tl) 81	Lead (Pb) 82	Bismuth (Bi) 83	Polonium (Po) 84	Astatine (At) 85		Radon (Rn) 86
Francium (Fr) 87	Radium (Ra) 88	Actinide Series 89–100						

some of them exist in large quantities and are commonly found throughout the world, e.g. oxygen, hydrogen and carbon, while others such as gold, uranium and radium are relatively rare, and some do not naturally occur on earth and are artificially created in nuclear physics equipment.

An atom of any element consists of a complex pattern of electrons that surround a positively charged nucleus. The electrons are assumed to travel in various orbits around the nucleus. The electrons each have a negative charge of 1.602×10^{-19} coulomb (known as the electronic charge e) and exist in just sufficient number to make the overall electrical charge of the atom equal to zero. The nucleus itself consists of a certain number A of particles known as nucleons. A is the mass number of the atom. There are two kinds of nucleons: *protons*, which each have a positive charge of e coulomb and *neutrons*, which have zero charge. The number of protons in a nucleus is known as the atomic number of the atom, symbol Z, and the number of neutrons is known as the neutron number N.

Hence $A = Z + N$.

The atomic number of each element is given in Table 1.1.

The difference between the atoms of the various elements lies in the number and arrangement of the electrons, protons and neutrons of which the atoms are composed. There is no difference between an electron in one element and another electron in any other element.

The Hydrogen Atom

The simplest atom is that of the element hydrogen and it consists merely of a single proton in the nucleus and a single electron in orbit around it (Fig. 1.1(a)). The helium atom is the next simplest and can be seen from Fig. 1.1(b) to consist of a nucleus (containing two protons and two neutrons) with two electrons orbiting around it.

For an electron to be able to move in a circular path around a nucleus, as shown in Fig. 1.1, it must have a force exerted on it pulling it towards the nucleus. This force is the electric attractive force exerted by the positive nucleus on the negative electron. Work must be done

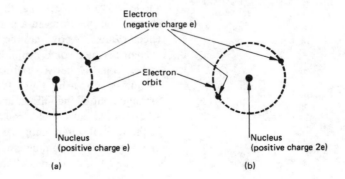

Fig. 1.1 The hydrogen and helium atoms

(a)   (b)

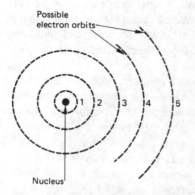

Fig. 1.2 Possible electron orbits in a hydrogen atom

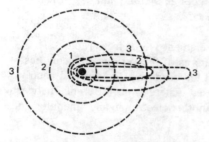

Fig. 1.3 Possible elliptical electron orbits in a hydrogen atom

in moving an electric charge through an electric field and so some work must have been done to move the electron from the nucleus to the orbit in which it is travelling. Thus an electron must possess a certain, discrete, quantity of energy before it can exist in an orbit around the nucleus. An electron can only exist in certain orbits of particular radii, and when in one of these orbits it must then possess the particular amount of energy associated with that orbit. An electron cannot occupy any orbit other than one of the orbits allowed as shown in Fig. 1.2. Normally the electron travels in the innermost orbit since this is the orbit of least energy, but if it is given, in some way, extra energy, such as heat, it may move to another orbit. An electron can only absorb the exact amount of energy required to raise its total energy to the energy value associated with another orbit; if given this amount of energy the electron will move to its new orbit and remain there until it loses some of its energy. The electron can only lose the exact amount of energy that will allow it to fall into a lower-energy orbit. This means that an electron can only absorb or lose energy in discrete amounts.

The simplified picture of the hydrogen atom given so far cannot account for all the observed phenomena, and it is necessary to extend the model by imagining that the electron can also move in elliptical orbits (Fig. 1.3). The number of elliptical orbits possible is equal to $n-1$, where n is the number of the basic circular orbits. The innermost orbit, $n = 1$, has no elliptical paths associated with it; the next orbit, $n = 2$, has a single elliptical path and so on. The circular orbits, numbers 1, 2, 3, etc. are said to form the K, L, M, etc. *shells*. The elliptical orbits are said to form sub-shells within these shells.

Other Atoms

The extra-nuclear make-up of the other, more complex, elements can be deduced with accuracy up to the element of atomic number 18 (argon), by adding one more electron for each element in turn, bearing in mind that the number, x, of electrons permitted in a particular shell is given by the expression $x = 2n^2$, where n is the order of the shell. The innermost shell can only contain 2×1^2 or 2 electrons, the next shell 2×2^2 or 8 electrons, the next shell 2×3^2 or 18 electrons, and so on. The electrons in a particular shell follow paths of different eccentricities. Above atomic number 18 some gaps appear in this system because some orbits in the N shell have lower energy than some orbits in the M shell and so they are filled first.

This is shown by Figs. 1.4(*a*) and (*b*), which give the nuclear arrangement of the silicon and germanium atoms respectively. The silicon atom has 14 electrons; two fill the innermost orbit, known as the K shell, eight complete the next, L, shell, leaving four further electrons in the incomplete outer, or M, shell. The germanium atom has its K, L and M shells completely filled and four electrons in its

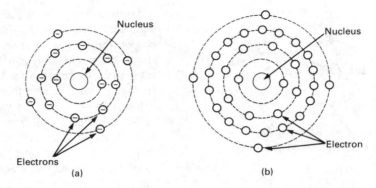

Fig. 1.4 (*a*) silicon and (*b*) germanium atoms

outer, or N, shell. In both cases there are four *valence* electrons in the outer shell.

The atoms of all elements in group III have three electrons that are not part of a completed, or closed, shell or sub-shell. Aluminium, for example, has 13 electrons, 10 of which completely fill the K and L shells; the M shell contains only three electrons and hence it is incomplete (since 18 electrons are necessary to fill it).

Indium is also in group III and it has 49 electrons, 28 of which complete the K, L and M shells. Of the remaining 21 electrons, 18 completely fill three of the four sub-shells of the N shell and the remaining three enter the O shell (leaving the fourth sub-shell of the N shell unfilled). For both aluminium and indium, therefore, the extra-nuclear structure consists of a number of tightly bound closed shells and sub-shells with three electrons outside and not so tightly bound to the nucleus.

The electronic structure of all the other atoms is also in the form of a number of closed shells and sub-shells with a number of electrons in orbit outside. The nucleus of an atom plus the closed shells and sub-shells of electrons can be considered to be a positively charged central core, the positive charge being equal to $e \times n$, where e is the electronic charge and n is the number of electrons outside the central core. The number of electrons outside this central core is equal to the number of the group in the Periodic Table of the Elements to which the atom belongs. Thus all atoms in group I may be represented by the sketch of Fig. 1.5(*a*), all atoms in group II by Fig. 1.5(*b*) and so on.

The outer electrons are known as valence electrons and they determine the chemical properties of the element.

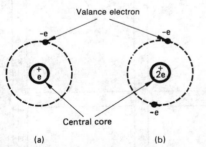

Fig. 1.5 Representation of (*a*) group I atoms and (*b*) group II atoms

Intrinsic Semiconductors

The two semiconductor materials used in the manufacture of semi-conductor devices, such as diodes, integrated circuits and transistors, are germanium and silicon. It can be seen from Table 1.1 that both these materials fall into group IV of the Periodic Table of the Elements. An atom of either substance may be represented by a central core

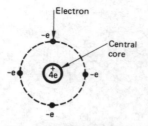

Fig. 1.6 Representation of germanium or silicon atom

of positive charge 4e surrounded by four orbiting electrons each of which has a negative charge of e (Fig. 1.6). In the remainder of this chapter the discussion of semiconductors will be with reference to silicon but it will apply equally well to germanium; any differences between the two materials will be mentioned at the appropriate places. In its solid state, silicon forms crystals of the diamond type, that is it forms a cubic lattice in which all the atoms (except those at the surface) are equidistant from their immediately neighbouring atoms. A study of crystal structures shows that the greatest number of atoms that can be neighbours to a particular atom at an equal distance away from that atom and yet be equidistant from one another is four. This means that each atom in a silicon crystal has four neighbouring atoms. In the crystal lattice each atom employs its four valence electrons to form a *covalent bond* with each of its four neighbouring atoms; each bond consisting of two electrons, one from each atom as shown in Fig. 1.7(a). Each pair of electrons traverses an orbit around both its parent atom and a neighbouring atom. Each atom is effectively provided with an extra four electrons and these are sufficient to complete its final sub-shell. To simplify the drawings in the remainder of this chapter, covalent bonding will be represented in the manner of Fig. 1.7(b).

If the temperature of the silicon is raised above absolute zero the crystal lattice is thermally excited and some of the valence electrons receive sufficient energy to enable them to break free from a covalent

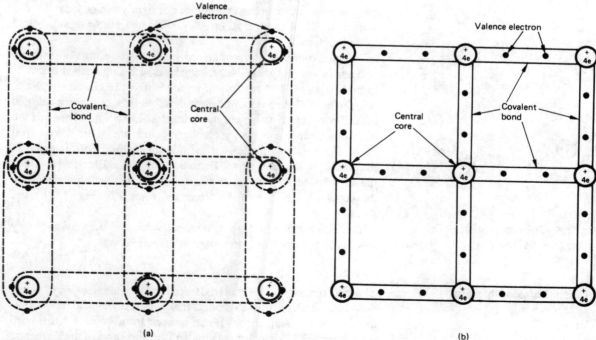

Fig. 1.7 Covalent bonding of atoms

bond. When this happens the liberated electrons wander at random in the crystal and are free to accept further energy from an applied electric field and to contribute to electrical conduction. With further increase in temperature more covalent bonds are broken, and the conductivity of the silicon is increased because of the increased number of free electrons. This means that a semiconductor material has a negative temperature coefficient of resistance.

When an electron escapes from a covalent bond it leaves behind it an 'absence of an electron' which, since it consists of a missing negative charge e, is equivalent to a positive charge of magnitude e. Such a positive charge is known as a *hole*. A hole exerts an attractive force on an electron and it can be filled by a nearby passing electron that has been previously liberated from another broken covalent bond. This process is known as recombination and it causes a continual loss of both holes and free electrons. At any given temperature the rate of recombination of holes and electrons is always equal to the rate of production of new holes and electrons so that the total number of free electrons and holes is constant.

When a covalent bond is broken it is said that a *hole-electron pair* has been created; both holes and electrons are known as *charge carriers*. The lifetime of a charge carrier is the time that elapses between its creation and its recombination with a charge carrier of the opposite sign.

Movement of Holes through the Lattice

Fig. 1.8 shows a part of a silicon crystal in which the breaking of covalent bonds by thermal agitation of the lattice is taking place. In Fig. 1.8(*a*) thermal agitation of the crystal lattice has caused a covalent bond to break and produce a hole-electron pair at point A. An instant later a second hole-electron pair is produced at point B and two electrons are free to wander in the lattice (Fig. 1.8(*b*)). In Fig. 1.8(*c*) electron 2 has wandered close enough to the first hole to be attracted by its electric field and recombination has occurred. The original hole has apparently moved from position A to position B, but at the same time another hole-electron pair has been created at point C. Finally, in Fig. 1.8(*d*), free electron 3 has travelled across the lattice and has recombined with the hole at point B and the effect is as though a hole has moved through the lattice from point A to point C.

The movement of both holes and electrons through the lattice is quite random but the holes appear to travel more slowly than do electrons. (This is because the movement of a hole in a particular direction actually consists of a series of discontinuous electron movements in the opposite direction.) If an electric field is set up in the crystal lattice the holes tend to *drift* in the direction of the field and the electrons to *drift* in the opposite direction. Thus conduction of current in a pure semiconductor, known as intrinsic conduction,

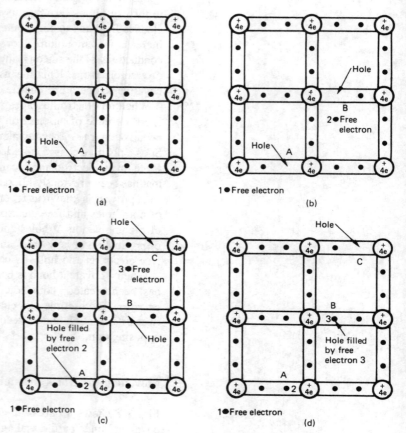

Fig. 1.8 The movement of holes through a crystal lattice
(a) Hole-electron pair created
(b) Second hole-electron pair created
(c) First hole disappeared and third hole-electron pair created
(d) Second hole disappeared

takes place (current flow is conventionally in the opposite direction to electron flow). Intrinsic conduction increases with increase in temperature at the approximate rate of 5% per degree Centigrade for germanium and 7% per °C for silicon.

Conductors and Insulators

A material acts as a conductor, or as an insulator, or as a semiconductor as a consequence of the way in which its atoms are bound together. Semiconductors employ covalent bonding as just explained. In a conductor, such as silver or copper, that is in group I of the Periodic Table, the single valence electron of an atom is easily detached and it is then available as a charge carrier. The atom that has lost an electron is then known as a positive *ion*. The bonding force which holds the atoms together to form a solid is the electric force of attraction that always exists between two charges of opposite sign; thus the positive ions and the negative electrons are attracted to one another. With increase in the temperature of the conductor the atoms tend to vibrate in their fixed positions and to obstruct the movement of the free electrons. This reduces the number of electrons passing from one point to another and so it causes the resistivity of the conductor to increase.

Some insulators, such as plastics and carbon, employ covalent bonding but with the valence electrons very firmly attached to their parent atoms. Hence very few charge carriers exist and the resistivity is very high. Other insulating materials employ ionic bonding; in this case some of the atoms lose valence electrons but these electrons move into orbit around another atom. As a result, some atoms lose an electron and become positive ions, while other atoms gain an electron and become negative ions. The electric force of attraction between positive and negative ions holds the atoms together to form a solid. Insulators in the first group have a negative temperature coefficient of resistance, and insulators in the second group have a positive temperature coefficient of resistance.

Extrinsic (Impurity) Semiconductors

If an extremely small, carefully controlled amount of an impurity element is introduced into a silicon crystal, each of the impurity atoms will take the place of one of the silicon atoms in the lattice. Since the number of impurity atoms is very much smaller than the number of silicon atoms (approximately 1 in 10^8), it is reasonable to assume that the lattice is essentially undisturbed and that each impurity atom is surrounded by four silicon atoms. In practice, the impurity elements are always substances in either group III or group V of the Periodic Table of the Elements that have either three or five valence electrons. Impurity elements typically employed are arsenic, antimony and phosphorus in group V, and indium, aluminium and gallium in group III. The process of introducing impurity atoms into a silicon crystal is called '*doping*' and a treated crystal is said to be 'doped'.

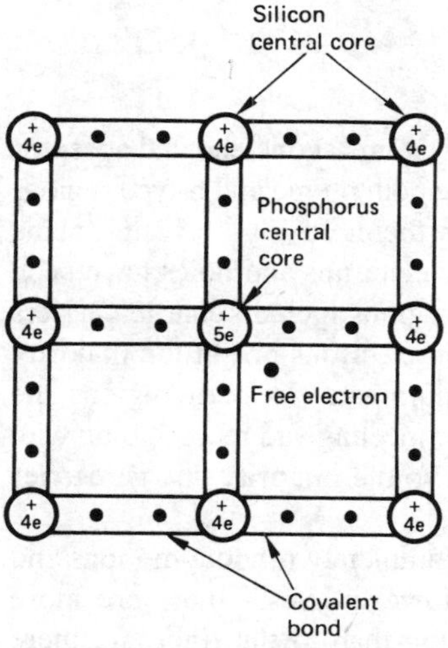

n-type Semiconductor

Suppose a silicon crystal has been doped with a small quantity of phosphorus, a substance having five valence electrons. Each phosphorus atom will set up covalent bonds with its four neighbouring silicon atoms but, since only four of its valence electrons are required for this purpose, a spare electron exists (Fig. 1.9). This surplus electron is not bound to its parent atom and is free to wander in the lattice.

A free electron is created in the silicon crystal lattice for each impurity atom introduced without the creation of corresponding holes. Hole-electron pairs are, however, still produced by thermal agitation of the lattice. The number of free electrons in the silicon crystal is much greater than the number of holes, this means that negative charges predominate and so the crystal is said to be *n-type*. Since each impurity atom donates a free electron to the crystal lattice, the impurity atoms are known as *donor* atoms.

Fig. 1.9 Lattice of n-type silicon crystal

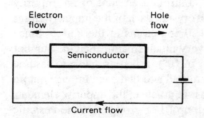

Fig. 1.10 Lattice of p-type silicon crystal

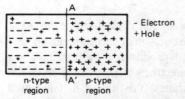

Fig. 1.11 Current flow in an extrinsic semiconductor

The p-n Junction

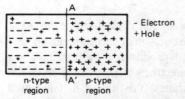

Fig. 1.12 The formation of a p-n junction

p-type Semiconductor

If, instead of phosphorus, a group III element such as boron is introduced into a silicon crystal, each boron atom will attempt to form a covalent bond with each of its four neighbouring silicon atoms. Boron, however, has only three valence electrons and so only three of the bonds can be completed (Fig. 1.10). One hole is introduced into the lattice for each impurity atom and is able to move about in the crystal in the same way as a hole that was produced by thermal agitation. In this case holes are in the majority and the material is known as *p-type*, while the impurity atoms are called *acceptor* atoms.

A crystal of n-type or p-type silicon is electrically neutral because each impurity atom introduced into the lattice is itself netural. In n-type material, electrons are the *majority charge carriers* and holes are the *minority charge carriers*. In p-type material the electrons are the minority charge carriers and holes are the majority charge carriers.

Current Flow

If a potential difference is maintained across an extrinsic semi-conductor (Fig. 1.11), a drift current will flow into the material at one end and out of the material at the other. The positively charged central cores cannot move from their positions in the crystal lattice and so the current flowing *into* the material can only consist of electrons flowing *out* and the current flowing *out* of the material is actually an inward flow of electrons.

If a silicon crystal is doped with donor atoms at one end and acceptor atoms at the other, the crystal will have both p-type and n-type regions and there will be a junction between them. In Fig. 1.12 the plane AA′ is the *p-n junction*; only the free electrons and holes have been shown to clarify the drawing. Both regions include charge carriers of either sign but in the n-type region electrons are in the majority and in the p-type region holes predominate. In both regions the probability of a minority charge carrier meeting and recombining with a majority charge carrier is high and so the minority charge carrier lifetime is short.

The free electrons and holes have completely random motions and they wander freely in the lattice. However, since there are more electrons to the left of the p-n junction than to the right and there are more holes to the right of the junction than to the left, *on average* more electrons cross the junction from left to right than cross from right to left, and more holes cross from right to left than cross from left to right. On average, therefore, the n-type region gains holes and loses electrons and the p-type region gains electrons and loses holes. This process is known as *diffusion* and it is the tendency for charge carriers to move away from areas of high density.

Since the n-type region loses negative charge carriers and gains positive charge carriers, and the p-type region loses positive charge carriers and gains negative charge carriers, the region just to the left of the junction becomes positively charged and the region just to the right of the junction becomes negatively charged. A hole passing into the n-type region, or an electron passing into the p-type region, becomes a minority charge carrier and will probably recombine with a carrier of opposite sign and disappear; however, one region has still lost a positive (or negative) charge and the other region has gained a positive (or negative) charge. The movement of holes and electrons across the junction constitutes a current and this is known as the *diffusion current*.

If the crystal was electrically neutral before diffusion took place it must still be neutral afterwards. Further, because both regions were also originally netural they must contain equal and opposite charges after diffusion. These charges have an attractive electric force between them and they are not able to diffuse away from the vicinity of the junction. The two charges are concentrated immediately adjacent to the junction and either side of it, and they produce a potential barrier across the junction. The polarity of the potential barrier is such as to oppose the further diffusion of majority charge carriers across the junction, but to aid the movement of minority charge carriers across the junction. This gives rise to a minority charge carrier current in the opposite direction to the diffusion current.

The difference in potential from one side of the junction to the other is called the *height of the potential barrier* and it is measured in volts. The height of the potential barrier attains such a value that the majority charge carrier (diffusion) and minority charge carrier currents are equal to one another and so the net current across the junction is zero. Any charge carriers entering the region on either side of the junction over which the barrier potential is effective are rapidly swept out of it, and hence this region is depleted of charge carriers.

The *depletion layer*, as it is called, is a region of relatively high resistivity and it is approximately 0.001 mm in width.

If an external source of e.m.f. is applied across the p-n junction, the equilibrium state of the junction is disturbed and the potential barrier is either increased or decreased according to the polarity of the external e.m.f. The silicon crystal consists of two regions of low resistivity separated by a region of high resistivity, the depletion layer (see Fig. 1.13), and the application of an e.m.f. across the crystal is effectively the same as placing it across the depletion layer.

The Forward-biased p-n Junction

If a battery is connected across the silicon crystal in the direction shown in Fig. 1.14, holes are repelled from the positive end of the crystal and are caused to drift towards the junction; and electrons are

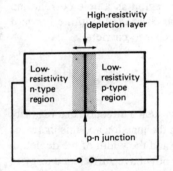

Fig. 1.13 The unbiased p-n junction

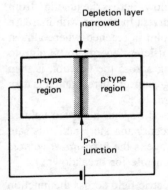

Fig. 1.14 The forward-biased p-n junction

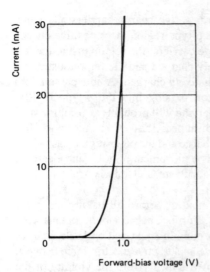

Fig. 1.15 The current-voltage characteristic of a forward-biased p-n junction

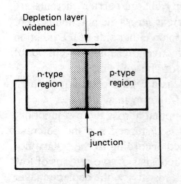

Fig. 1.16 The reverse-biased p-n junction

repelled from the negative end of the crystal and also drift towards the junction. This drift of holes and electrons towards the junction reduces both the width of the depletion layer and the height of the potential barrier, and the junction is said to be *forward biased*. The reduction in the height of the potential barrier allows majority charge carriers of lower energy to cross the junction and, since the minority charge carrier current remains constant, there is a net majority charge carrier current across the junction from the p-type region to the n-type region. This current increases very rapidly with increase in the forward bias voltage as can be seen from the typical current-voltage characteristic shown in Fig. 1.15.

The holes drifting through the p-type region towards the p-n junction may be considered to have been injected by the positive terminal of the battery. Some of these holes may recombine with electrons diffusing across the junction in the other direction and so the hole current across the junction is slightly less than the injected hole current. After they have passed across the junction the holes recombine with the excess electrons in the n-type region. Similarly, the negative battery terminal injects electrons into the n-type region and most of these electrons cross the junction. The total current is the sum of the electron and hole currents and is constant throughout the silicon crystal. The current enters the p-type region as a hole current and leaves the n-type region as an electron current, i.e. in the forward direction current flow is by majority charge carriers.

The Reverse-biased p-n Junction

Fig. 1.16 shows a p-n junction biased in such a direction that majority charge carriers are attracted away from the junction and this increases both the height of the potential barrier and the width of the depletion layer. Fewer majority charge carriers now have sufficient energy to be able to surmount the potential barrier and so the majority charge carrier current decreases. The minority charge carrier current has remained constant and so a net current flows across the junction from n-type region to p-type region. This current increases with increase in the reverse-bias voltage until the point is reached where almost no majority charge carriers possess sufficient energy to be able to cross the junction. The current flowing across the junction is then constant and equal to the minority charge carrier current and it is then known as the *reverse saturation current*.

If the reverse-bias voltage is increased beyond a certain value a rapid increase in the reverse current occurs and the junction is said to have broken down; this critical voltage is the *breakdown voltage* of the junction. Two effects are responsible for breakdown:

(a) the *Zener effect* in which the electric field across the junction is strong enough to break some of the covalent bonds and

(b) the *avalanche effect* in which charge carriers are accelerated to such an extent that they are able to break covalent bonds by collision.

A typical current-voltage characteristic for a reverse-biased p-n junction is shown in Fig. 1.17. If the temperature of a p-n junction is increased the number of minority charge carriers that are generated and are able to cross the junction will also increase. This will increase the reverse saturation current. The reverse breakdown current may damage a diode unless it is limited to a safe value by a series resistance. Manufacturers of semiconductor devices recommend maximum junction temperatures which should not be exceeded. For most devices these limits are:

germanium 90°C
silicon 150°C (plastic case)
       200°C (metal case)

The Capacitance of a p-n Junction

When a p-n junction is reverse-biased, the depletion layer is a high-resistance region that has low-resistance regions on either side and so it acts as though it were a parallel-plate capacitor, the capacitance of which is a function of the magnitude of the applied bias voltage. A p-n junction can be made with the transition from the p-type region to the n-type region either abrupt or gradual. For an abrupt junction the depletion capacitance is proportional to the square root of the bias voltage, and for a gradual junction it is proportional to the cube root of the reverse bias voltage.

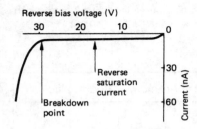

Fig. 1.17 The current-voltage characteristic of a reverse-biased p-n junction

Exercises

1.1 An atom of indium has 49 electrons. How many of these are (a) in the innermost orbit, (b) in the outermost orbit?

1.2 How many electrons has a copper atom? How many of these are valence electrons? How many protons are there in its nucleus?

1.3 Arsenic has atomic number 33 and is in group V of the Periodic Table of the Elements. How many electrons are there in (a) the innermost orbit, (b) the outermost orbit?

1.4 Give two examples of a conductor, an insulator and a semiconductor. State the effect of temperature change on the resistivity of each.

1.5 The number x of electrons allowed in a particular shell is given by $x = 2n^2$. Calculate the maximum number of electrons in each of the first three shells.

1.6 The charge of an electron is 1.602×10^{-19} C. Calculate the charge of the central core of a silicon atom.

1.7 Explain how a hole moves through a silicon crystal.

1.8 Silicon has a resistivity of 0.5 Ω m at 27°C. Calculate its resistivity when the temperature is 30°C.

1.9 Explain with the aid of a sketch how silicon atoms are held together by means of covalent bonds.

1.10 Draw an n-type region with a battery connected across it. Indicate (*a*) the direction in which electrons move, (*b*) the direction in which holes move and (*c*) the direction of current flow out of the battery.

1.11 For Table 1.2, tick as appropriate *either* the donor *or* the acceptor box beneath each element.

Table 1.2

Element	Aluminium	Arsenic	Antimony	Boron	Indium
Donor					
Acceptor					

1.12 An n-type region contains more free electrons than holes. Why, therefore, is the semiconductor electrically neutral?

1.13 An intrinsic semiconductor is doped with trivalent atoms. What charge has the resulting extrinsic material? Briefly explain your answer.

1.14 What is the polarity of the potential barrier? Should it be increased or decreased in order for a majority charge carrier current to flow? What polarity-bias voltage is needed to achieve this?

1.15 A p-n junction is forward biased. Draw a sketch to show (*a*) the majority charge carrier current, (*b*) the minority charge carrier current, (*c*) the hole current and (*d*) the battery current.

1.16 Assuming the barrier potential to be 0.6 V calculate the current flowing in Fig. 1.18(*a*).

1.17 Assuming the barrier potential to be 0.62 V calculate the current flowing in Fig. 1.18(*b*).

1.18 What is a depletion layer and how is it formed? Is it a region of high or of low resistivity? Is the width of a depletion layer increased or decreased by a reverse-bias voltage? Will this assist in or oppose the flow of minority charge carriers across the junction?

1.19 Explain why a p-n junction has self-capacitance. Is this capacitance increased or decreased by an increase in the reverse-bias voltage?

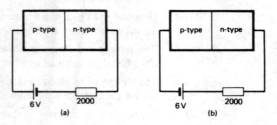

Fig. 1.18

2 Semiconductor Diodes

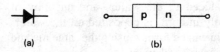

(a)

(b)

Fig. 2.1 The basic semiconductor diode

The semiconductor diode is a device that has a high resistance to the flow of current in one direction and a low resistance in the other. The symbol for a semiconductor diode is shown in Fig. 2.1(a). The direction in which the diode offers little opposition to current flow is indicated by the arrowhead. The diode is widely employed in electronic circuitry for many different purposes and it consists essentially of a p-n junction formed in either a silicon or a germanium crystal (Fig. 2.1(b)).

Germanium and silicon for use in the manufacture of semiconductor diodes must first be purified until an impurity concentration of less than 1 part in 10^{10} is achieved. The wanted impurity atoms, donors and/or acceptors, are then added in the required amounts and the material is made into a single crystal.

Construction

A p-n junction may be formed in a number of different ways but two basic techniques are generally employed, either singly or in combination. An example of the first method is outlined in Fig. 2.2 and consists of alloying an indium pellet on to an n-type germanium wafer.

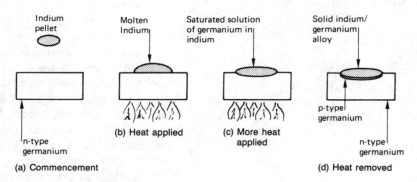

Fig. 2.2 The 'alloying' method of forming a p-n junction

To make the n-type germanium wafer some intrinsic germanium and a small amount of a pentavalent impurity are melted in a crucible in a vacuum, and a seed crystal is lowered into the melt to a depth of a few millimetres. The temperature of the molten germanium is just above the melting point of the seed crystal, and the few millimetres of the seed immersed in the melt also melts. The seed crystal is rotated at a constant velocity and at the same time it is slowly withdrawn from the melt, to form an n-type crystal. By careful control of the process the required impurity concentration can be achieved.

A pellet of indium is placed on the n-type germanium wafer and is then heated to a temperature above the melting point of indium but below the melting point of germanium. The indium melts and dissolves the germanium until a saturated solution of germanium in indium is obtained. The wafer is then slowly cooled and in the cooling a region of p-type germanium is produced in the wafer, and an alloy of germanium and indium (mainly indium) is deposited on the wafer.

A silicon alloyed p-n junction can be formed using the same method but using aluminium as the acceptor element.

The second method of producing a p-n junction to be considered here is diffusion and it is outlined in Fig. 2.3. The p-type germanium wafer is heated to a temperature very nearly equal to the melting point of germanium, and it is surrounded by the donor element antimony in gaseous form. The antimony atoms will diffuse into the germanium to produce an n-type region. If an n-type germanium crystal is used gallium is employed, in gaseous form, as the acceptor element to produce a p-type region in the crystal. When a silicon device is to be manufactured, boron is used as the acceptor element and phosphorus as the donor element.

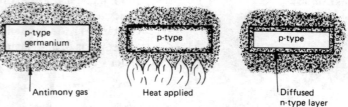

Fig. 2.3 The 'diffusion' method of forming a p-n junction

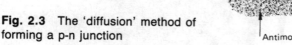

A junction diode consists of a crystal having both p-type and n-type regions. Junction diodes are made from either germanium or silicon, the former having the advantage of a lower forward resistance and the latter the advantages of a higher breakdown voltage and a lower reverse saturation current. Connection to the junction is made by wires fixed to each of the two regions. The complete device is usually enclosed in a hermetically sealed container to prevent the entry of moisture (see Fig. 2.4(a)).

The silicon planar diode is manufactured using the diffusion method, and a typical construction is shown in Fig. 2.4(b).

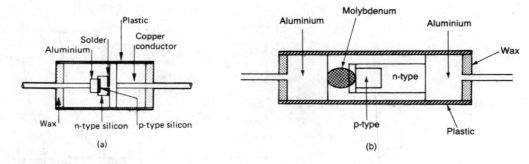

Fig. 2.4 Construction of (a) a silicon junction diode and (b) a silicon planar diode

Semiconductor Diode Current-Voltage Characteristics

The current-voltage characteristic of a semiconductor diode is a graph of the current flowing in the device plotted against the voltage applied across it.

It can be measured with the aid of the circuit arrangement of Fig. 2.5. With the switch in the position shown, the diode is reverse-biased; to forward bias the diode, the switch is thrown to its other position in order to reverse the polarity of the applied voltage. For each position of the switch, the applied voltage is increased from zero in a number of steps and the current flowing at each step is noted. The noted current values are then plotted to a base of voltage.

Typical current-voltage characteristics for signal silicon and germanium diodes are shown in Fig. 2.6. Note that the forward current is in milliamps while the reverse current is in microamps. A power diode would have a forward current measured in amps and a reverse current of a few milliamps.

The forward current does not increase to any noticeable extent until the forward bias voltage is greater than about 0.6 V for the silicon diode and about 0.2 V for the germanium diode. The other features of importance are (i) the reverse saturation current, and (ii) the reverse breakdown voltage (not shown). The a.c. resistance r_{ac} of a diode at a particular d.c. voltage is equal to the reciprocal of the slope of the characteristic at that point, that is

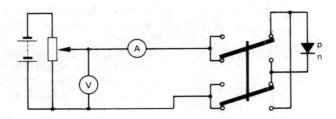

Fig. 2.5 Circuit for measuring the current-voltage characteristic of a semiconductor diode

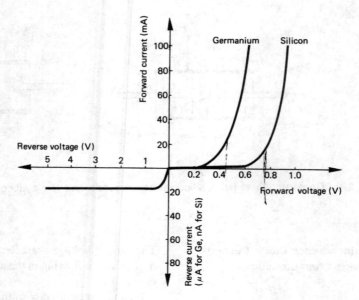

Fig. 2.6 Current-voltage
characteristics for a silicon diode and
a germanium diode

$$r_{ac} = \frac{\text{Change in voltage}}{\text{Resulting change in current}} = \frac{\delta V}{\delta I} \text{ ohm} \qquad (2.1)$$

[*Note* The Greek letter δ (delta) means 'a change of' wherever it
appears in formulae. So, δt is a change of time. Generally it indicates
a small-scale change.]

At any point along the characteristic the ratio (voltage applied)/
(current flowing) is a measure of the d.c. resistance r_{DC} of the diode
for that voltage. If the characteristic is linear both the a.c. resistance
and the d.c. resistance of the diode will be constant quantities but
should the characteristic be non-linear, the resistances will vary with
the point of measurement.

Example 2.1

Calculate the a.c. resistance of the semiconductor diode whose characteristic
is shown in Fig. 2.7 at the point +1 V.

Solution

The a.c. resistance r_{ac} is equal to $\delta V/\delta I$ and to find the forward resistance
at the point $V = 1$ V it is necessary to select two equidistant points either
side of this voltage and then, by projection to and from the characteristic,
find the corresponding values of current.

Two points 0.2 V either side of +1 V have been selected, hence
$\delta V = 0.4$ V.

Projection upwards from these points to the curve and then from the curve
to the current axis, as shown by the dotted lines, shows that the corresponding values of current are 15.5 mA and 5.5 mA, i.e. $\delta I = 10$ mA.

Therefore $r_{ac} = 0.4/(10 \times 10^{-3}) = 40$ ohms   (*Ans.*)

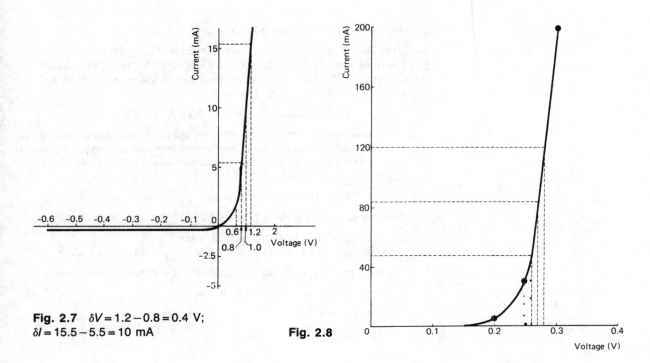

Fig. 2.7 $\delta V = 1.2 - 0.8 = 0.4$ V;
$\delta I = 15.5 - 5.5 = 10$ mA

Fig. 2.8

The slope of the reverse saturation current curve is very small and cannot be measured from the characteristic; this means that the reverse a.c. resistance of the diode is high, usually some thousands of ohms.

Example 2.2

The current-voltage characteristic of a semiconductor diode is given by Table 2.1.

Table 2.1

Forward voltage (V)	0.05	0.10	0.15	0.20	0.25	0.30
Forward current (mA)	0.2	0.4	0.6	4.0	30	200

Plot the characteristic and use it to determine (i) the d.c. resistance and (ii) the a.c. resistance of the diode when the forward-bias voltage is (a) $V = 0.27$ volts and (b) 0.18 volts.

Solution

The current-voltage characteristic of the diode is shown plotted in Fig. 2.8.

(a)(i) The d.c. resistance of the diode at the point $V = 0.27$ V is found by drawing a line upwards from the voltage axis to the characteristic and then projecting on to the current axis. The corresponding d.c. current value is 84 mA and therefore

$$r_{DC} = V/I = 0.27/(84 \times 10^{-3}) = 3.21 \ \Omega \quad (Ans.)$$

(ii) The a.c. resistance of the diode is determined using the method of the preceding example. Points 0.01 V either side of 0.27 V have been selected so that $\delta V = 0.02$ V. The corresponding current values are 120 mA and 48 mA; hence $\delta I = 72$ mA. Therefore

$$r_{ac} = \delta V/\delta I = 0.02/(72 \times 10^{-3}) = 0.28 \ \Omega \quad (Ans.)$$

[It is important that small increments of voltage are chosen when calculating r_{ac}, otherwise considerable error may occur. Suppose, for example, that points 0.03 V either side of $V = 0.27$ V had been selected. Then $\delta V = 0.06$ V and $\delta I = 178$ mA giving

$$r_{ac} = 0.06/(178 \times 10^{-3}) \text{ or } 0.34 \ \Omega$$

This is a percentage error of $[(0.34 - 0.28)/0.28] \times 100$ or 21.43%.]

(b) (i) $r_{DC} = 0.2/(4 \times 10^{-3}) = 50 \ \Omega \quad (Ans.)$
(ii) $r_{ac} = (0.19 - 0.17)/[(2.7 - 1.3) \times 10^{-3}] = 14.3 \ \Omega \quad (Ans.)$

A good approximation to the a.c. resistance of a forward-biased diode is

$$r_{ac} = 25/I_F \quad \text{where } I_F \text{ is in milliamps} \tag{2.2}$$

Thus, in example 2.2(a)(ii),

$$r_{ac} = 25/84 = 0.3 \ \Omega \quad (Ans.)$$

Forward Voltage Drop

When a voltage is applied across a semiconductor diode with the polarity required to forward bias its p-n junction, the barrier potential is reduced, which allows more majority charge carriers to cross the junction. Since the minority charge carrier current is unaffected, a net current flows across the junction. As the forward-bias voltage is increased, the current is very small at first but it increases rapidly once the voltage has exceeded a particular threshold value. For a silicon diode this threshold voltage is approximately 0.6 V, but it is only about 0.2 V for a germanium diode. This is clearly shown by the typical characteristics shown in Fig. 2.6.

Forward Current

For each type of diode, a maximum forward average current $I_{F(AV)}$ and a maximum repetitive forward current I_{FRM} are quoted by the manufacturer.

Reverse Saturation Current

When a reverse-bias voltage is applied to a semiconductor diode, the barrier potential is increased and then fewer majority charge carriers have sufficient energy to cross the junction. With increase in the reverse-bias voltage, the point is reached where the current consists almost entirely of minority charge carriers. The current flowing then becomes more or less constant and is known as the *reverse saturation current*. The reverse saturation current in a germanium diode is very much greater than the reverse saturation current of a silicon diode of comparable maximum forward current rating. Thus, for the smaller types of diode the reverse saturation current would be a few microamperes in a germanium diode but only a few nanoamperes in a silicon diode.

If the temperature of the p-n junction is increased, further hole electron pairs will be produced and the reverse saturation current will become larger.

Breakdown Voltage

If the reverse-bias voltage applied to a diode is steadily increased, the current will remain at an approximately constant value until a point is reached where a sudden and large increase in current takes place (Fig. 2.9). In the breakdown region the reverse resistance of the diode will be low. This large increase in the reverse current will dissipate power within the diode and it may lead to the destruction of the device. It is necessary, therefore, to ensure that the diode is not driven into its breakdown region. An arbitrary voltage rating is determined and quoted by the manufacturer for each type of diode which, if not exceeded, will ensure the satisfactory working of the device. The *peak reverse repetitive voltage* varies considerably with the type of diode and may easily be a few hundreds of volts.

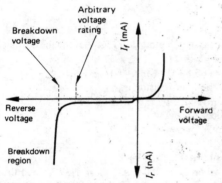

Fig. 2.9 Voltage breakdown in a semiconductor diode

Example 2.3

A low-power diode has the current-voltage characteristic given by the figures in Table 2.2.

Table 2.2

Forward voltage (V)	0	0.7	0.8	0.9	1.0	1.1	1.2	1.3	1.4
Forward current (mA)	0	1	5	28	65	120	175	240	330

The diode is connected in the circuit shown in Fig. 2.10(*a*). Calculate (*a*) the current flowing and (*b*) the value of the load resistor R_L, (*c*) the power dissipated in both the diode and the load resistor, (*d*) the d.c. resistance of the diode and (*e*) the smallest allowable load resistance if the maximum forward current is 300 mA.

Solution
The forward voltage drop across the diode is $9 - 7.9 = 1.1$ V.

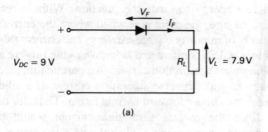

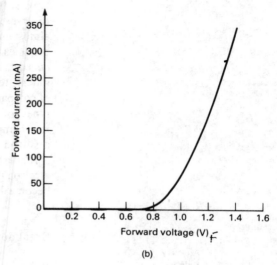

Fig. 2.10

(a) From the diode characteristic shown plotted in Fig. 2.10(b) the current flowing in the diode is 120 mA (*Ans.*)

(b) $R_L = 7.9/(120 \times 10^{-3}) = 65.83\ \Omega$ (*Ans.*) $V < IR$ $R = V/I$

(c) The power dissipated in the load resistor is
$$P_L = 0.12^2 \times 65.83 = 948\ \text{mW} \quad (Ans.)$$ $P = I^2 R$

The power dissipated in the diode is
$$P_D = 0.12 \times 1.1 = 132\ \text{mW} \quad (Ans.)$$ $P = VI$

(d) $r_{DC} = 1.1/0.12 = 9.17\ \Omega$ (*Ans.*)

[*Note* $V_L = (9 \times 65.83)/(9.17 + 65.83) = 7.9$ V]

(e) From the characteristic, when $I_F = 300$ mA $V_F = 1.37$ V, and the smallest possible value for the load resistance is $R_L = 7.63/0.3 = 25.4\ \Omega$ (*Ans.*)

Diode Load Lines

The current flowing through a diode in series with a resistance, and the voltage dropped across it, can be determined with the aid of a *load line* drawn on the diode's current-voltage characteristic. Referring to Fig. 2.10(a), the voltage across the diode is equal to the applied voltage minus the voltage dropped across the load resistor. Thus,

$$V_{DC} = V_F + V_L = V_F + I_F R_L \tag{2.3}$$

$$I_F = V_{DC}/R_L - V_F/R_L \tag{2.4}$$

This is the equation of a straight line and so only two points are required on the diode characteristic.

Point A: When $I_F = 0$ $V_F = V_{DC}$.

Point B: When $V_F = 0$ $V_{DC} = I_F R_L$, or $I_F = V_{DC}/R_L$.

The two points are shown plotted on a diode characteristic in Fig. 2.11. The load line is drawn to join the two points and it has a slope equal to the reciprocal of the load resistance. Both equation (2.3) and the diode current-voltage characteristic must be simultaneously

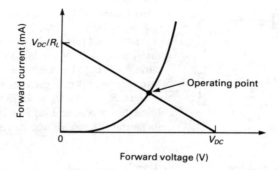

Fig. 2.11 D.C. load line on a diode
current-voltage characteristic

satisfied and this is possible only at the point of intersection of the
d.c. load line and the I/V curve. The point of intersection gives the
quiescent, steady, values of the current and voltage in, and across,
the diode and it is known as the *operating point*.

Example 2.4

The diode whose characteristics are given by Table 2.2 is used in the circuit
of Fig. 2.10(a). (a) Use a d.c. load line to determine the current flowing
in the circuit when the applied d.c. voltage is 5 V and the load resistance
is 100 Ω. (b) Calculate the voltage across the load resistance. (c) Find the
value of load resistance that will allow a current of 155 mA to flow in the
circuit.

Solution

The d.c. load line should be drawn between the points $I_F = 0$, $V_F = V_{DC}$
= 5 V and $V_F = 0$, $I_F = V_{DC}/R_L = 5/100 = 50$ mA. This is shown by Fig.
2.12. The operating point has been marked P.
  (a) Projecting from the operating point to the current axis gives
$$I_F = 40 \text{ mA} \quad (Ans.)$$

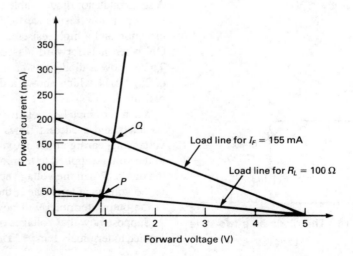

Fig. 2.12

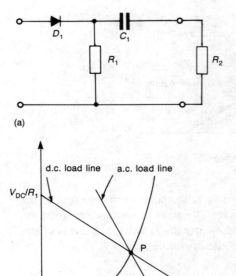

(a)

(b)

Fig. 2.13 (a) Diode with different a.c. and d.c. loads and (b) diode a.c. and d.c. load lines

The Semiconductor Diode as a Switch

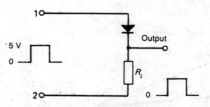

Fig. 2.14 The diode as a two-state device

(b) Projecting from the operating point to the voltage axis gives
$$V_F = 0.92 \text{ V} \quad (Ans.)$$
(c) The load line for the new value of load resistance must pass through the point $I_F = 0 \quad V_F = 5$ V and the point Q where a projection from the current axis at 155 mA cuts the curve. This is also shown by Fig. 2.12. The load resistance is then given by the reciprocal of the slope of the load line.
Thus, $R_L = \delta V_F/\delta I_F = 2/(78 \times 10^{-3}) = 25.6 \ \Omega \quad (Ans.)$
[Note that since the load line is linear it does not matter how large the increments of voltage and current that are chosen happen to be.]

A.C. Load Lines

Very often a diode circuit will have both d.c. and a.c. voltages applied to it and the load resistance into which the diode works may be different for both voltages. An example of differing a.c. and d.c. loads is shown by Fig. 2.13. For the d.c. operation of the circuit the load resistance is merely the resistor R_1 and so the d.c. load line is drawn between the points $V_F = V_{DC}$ and $I_F = V_{DC}/R_1$. The operating point of the circuit is, as before, given by the intersection of the d.c. load line and the I/V curve and has been labelled as P. The effective load resistance for the diode for a.c. signals is the total resistance of R_1 and R_2 in parallel, i.e. $R_1R_2/(R_1 + R_2)$. The a.c. load line must be drawn through the operating point P with a slope equal to $-(R_1 + R_2)/R_1R_2$.

The requirements for an electronic switch are:

(a) Can be turned ON and OFF by the application of a voltage
(b) Switches between its two states very rapidly
(c) Has a very high OFF resistance and a low ON resistance
(d) Dissipates very little power when either ON or OFF.

A semiconductor diode is able to operate as a two-state device because it offers a low resistance to the flow of an electric current in one direction and a high resistance in the other. The diode is said to be ON when it is forward biased and OFF when it is reverse biased. To see how a diode acts as a two-state device consider the circuit of Fig. 2.14 which shows a diode connected in series with a load resistor R_L.

When terminal 1 is positive with respect to terminal 2, the diode conducts and a current flows to develop a voltage across R_L. The voltage appearing at the output terminals of the circuit will be equal to the voltage applied to terminal 1 minus the voltage dropped across the diode. When the voltage applied to terminal 2 is positive relative to the voltage at terminal 1, the diode will not conduct. The voltage at the output terminal will now be equal to the voltage at terminal 2. Suppose now that voltages of the same magnitude and polarity are applied to terminals 1 and 2. The diode will not conduct and the output

Table 2.3 Voltage table

Terminal 1	Terminal 2	Output Terminal
Positive	Negative	Positive
Negative	Positive	Positive
Positive	Positive	Positive
Negative	Negative	Negative

voltage will be the same as the common value of the input voltages. The action of the circuit can be expressed by a *voltage table* such as Table 2.3

When a diode is employed as a switch the time it takes to switch from ON to OFF or from OFF to ON will be of importance. The time taken for a diode to turn ON is always small enough to be neglected but the turn-OFF time is larger and it provides a limiting factor to the maximum frequency at which the diode may be switched.

Fig. 2.15 shows a rectangular voltage that is applied to a diode. When the diode is forward biased the diode current flows more or less instantaneously. When the polarity of the applied voltage is reversed to turn the diode OFF, the diode does *not* immediately turn OFF, with its current falling to its reverse saturation value. Instead, the diode conducts current in the reverse direction to produce an initial surge of current I_R that is equal to, or very nearly equal to, the forward current I_F. Then the reverse current falls to its reverse saturation value. The time that elapses between the diode voltage changing its polarity and the diode current reaching 10% of its maximum reverse value is known as the *reverse recovery time t_{rr}* and it is usually measured in nanoseconds.

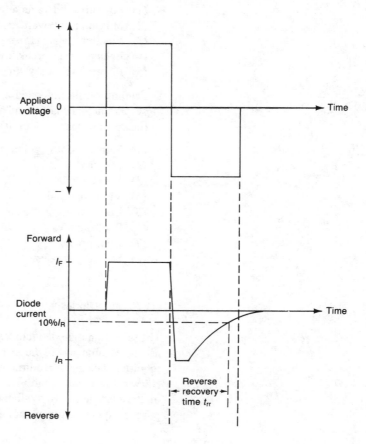

Fig. 2.15 Switching a diode

The excess reverse current occurs because at the instant the voltage applied to the diode changes its polarity there are a large number of charge carriers crossing the p-n junction and these must all be removed before the diode current is merely provided by the minority charge carriers.

To avoid having a surge of reverse current the reverse recovery time must be several (usually 10) times smaller than the periodic time of the applied waveform.

Example 2.5

Calculate the maximum frequency at which a diode having a reverse recovery time $t_{rr} = 5$ ns can be switched.

Solution

$$T = 10\, t_{rr} = 50 \text{ ns}.$$

Therefore $f_{max} = 1/(50 \times 10^{-9}) = 20$ MHz (*Ans.*)

Types of Diode and their Application

The important parameters of semiconductor diodes are

(1) Forward and reverse a.c. resistances
(2) Maximum forward current
(3) Junction capacitance
(4) Behaviour in breakdown region
(5) Reverse recovery time.

Depending on the intended application of a diode, one or more of these parameters may be of prime importance.

The main types of diode used in modern electronic circuitry are

(1) Signal diodes (which include switching diodes)
(2) Power diodes
(3) Zener diodes
(4) Varactor diodes
(5) Light-emitting diodes
(6) Photo-diodes
(7) PIN diodes
(8) Schottky diodes.

(1) Signal Diodes

The term signal diode includes all diodes that have been designed for use in circuits where large current and/or voltage ratings are not required. The usual requirements are for a large (reverse resistance)/ (forward resistance) ratio and minimum junction capacitance. Some of the commercially available signal diodes are listed as general-purpose types while others are best suited to a particular circuit application, e.g. as a detector of radio waves, or as an electronic switch in logic circuitry. The maximum reverse voltage that the diode is likely to be called on to handle is usually not very high, and neither is the maximum forward current. Most types of signal diode have a peak reverse repetitive voltage in the range 30 V to 150 V and a maximum forward current somewhere between 40 and 250 mA, but higher values are readily available.

Arrays of signal diodes are available from various manufacturers. Four or more diodes are provided within a single dual-in-line package (page 110).

(2) Power Diodes

Power diodes are most often employed for the conversion of alternating current into direct current, i.e. used as rectifiers. The important power diode parameters are the peak voltage, the maximum forward current and the resistance ratio. The peak reverse voltage is likely to be somewhere between 50 V and 1000 V with a maximum forward current of perhaps 30 A. The forward resistance must be as low as possible to avoid considerable voltage drop across the diode when the large forward current flows; this resistance is usually not very much more than an ohm or two. Power diodes are slow to switch.

Power diodes are often used in full-wave bridge rectifier circuits, and four diodes in bridge form can be obtained in a single package.

(3) Zener Diodes

The large reverse current which flows when the breakdown voltage of a diode is exceeded need not necessarily result in damage to the device. A Zener diode is fabricated in a way which allows it to be operated in the breakdown region without damage, provided the current is restricted by external resistance to a safe value. The large current at breakdown is brought about by two factors, known as the Zener and the avalanche effects. At voltages up to about 5 V the electric field near to the junction is strong enough to pull electrons out of the covalent bonds holding the atoms together. Extra hole-electron pairs are produced and these are available to augment the reverse current. This is known as the *Zener effect*.

The *avalanche effect* occurs if the reverse-bias voltage is made larger than 5 V or so. The velocity with which the charge carriers move through the crystal lattice is increased to such an extent that they attain sufficient kinetic energy to *ionize* atoms by collision. An atom is said to have been ionized when one of its electrons has been removed. The extra charge carriers thus produced travel through the crystal lattice and may also collide with other atoms to produce even more

Fig. 2.16 Zener diode symbol

carriers by ionization. In this way the number of charge carriers, and hence the reverse current, is rapidly increased. In the forward direction a Zener diode behaves just like a silicon signal diode. The symbol for a Zener diode is shown in Fig. 2.16.

Zener diodes are available in a number of standardized *reference voltages*. For example, it is possible to obtain a Zener diode with a reference (breakdown) voltage of 8.2 V. An alternative name for the device is the *voltage reference diode*.

Example 2.6

A Zener diode is advertised as having a breakdown voltage of 20 V with a maximum power dissipation of 400 mW.

What is the maximum current the diode should be allowed to handle?

Solution

$$I = P/V = 0.4/20 = 20 \text{ mA} \quad (Ans.)$$

A typical Zener diode characteristic is shown in Fig. 2.17. The forward current is the same as that of a small-signal diode. The a.c. resistance of the device when it is the breakdown mode is given by the ratio $\delta V_Z/\delta I_Z$.

Zener diode reference voltages have a tolerance of about 5% so that the actual voltage will vary somewhat either side of the nominal value. Typical reference voltages are quoted on page 35 but a typical one is 8.2 V. The actual voltage of such a device may vary from 7.79 volts to 8.61 volts. Variations in the input voltage will cause the Zener current to vary and this will bring a small further change in the reference voltage. A Zener diode is often used as a voltage regulator in a power supply (see page 154) or as a voltage reference. Two examples of the latter use are shown by Fig. 2.18. In (*a*) the Zener diode is used for over-voltage protection; if the input voltage should

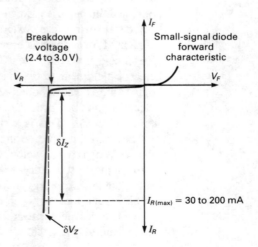

Fig. 2.17 Zener diode current-voltage characteristic

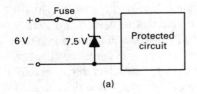

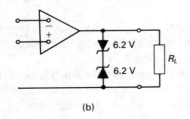

Fig. 2.18 Use of a Zener diode to (*a*) protect a circuit from a high input voltage and (*b*) prevent the output voltage of an amplifier rising above a set value

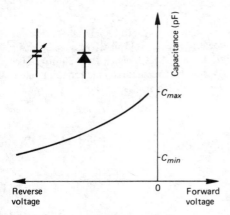

Fig. 2.19 Varactor diode characteristics

rise to 7.5 V the Zener diode will conduct current and this will cause the fuse to blow. In (*b*) the output of an op-amp is prevented from becoming higher than ±6.2 V by two back-to-back connected Zener diodes.

(4) Varactor Diodes

A p-n junction is a region of high resistivity sandwiched between two regions of relatively low resistivity. Such a junction therefore possesses capacitance, the magnitude of which is given by

$$C = \frac{\epsilon A}{W} \tag{2.5}$$

where ϵ is the permittivity of the semiconductor material, A is the area of the junction, and W is the width of the depletion layer. W is not a constant quantity but, instead, varies with the magnitude and the polarity of the voltage applied across the junction.

Most semiconductor diodes are manufactured in such a way that their junction capacitance is minimized, but a varactor diode has been designed to have a particular range of capacitance values.

The varactor diode is operated with a reverse-bias voltage and then its junction capacitance is inversely proportional to the square root of the bias voltage V, i.e.

$$C = \frac{K}{\sqrt{V}} \tag{2.6}$$

Fig. 2.19 shows graphically how the capacitance of a varactor diode varies with the reverse-bias voltage, and it also shows the symbol for a varactor diode. Typically, the capacitance variation might be 2−12 pF, or 20−28 pF, or perhaps 27−72 pF.

Example 2.7

A varactor diode has a capacitance of 5 pF when the reverse-bias voltage applied across it is 4 V. Determine the diode capacitance if the bias voltage is increased to 6 V.

Solution

From equation (2.6) $5 = \dfrac{K}{\sqrt{4}}$ i.e. $K = 10$ pF $\sqrt{V}$

Therefore, when the voltage has increased to 6 V,

$C = 10/\sqrt{6} = 4.082$ pF   (*Ans.*)

(5) Light-emitting Diodes

A light-emitting diode (LED) is a junction diode that emits visible light energy whenever it passes a forward current. The voltage

Table 2.4

Material	Colour	Forward voltage (V)
GaAsP	Red	1.6
GaP	Green	2.1
GaAsP	Yellow	2.0
GaAsP	Orange	1.8
SiC	Blue	3.0

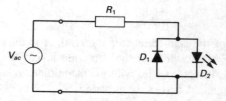

Fig. 2.20 Protecting a LED from a reverse voltage

required to turn the LED ON and the colour of the visible light are determined by the semiconductor materials that are used. The materials employed are gallium phosphide (GaP), gallium arsenide phosphide (GaAsP) and silicon carbide (SiC). The forward current must not be allowed to exceed a safe figure quoted by the manufacturer, generally some 20 to 60 mA. Table 2.4 shows the colour of the light emitted by each kind of LED and the forward voltage drop when the current flowing is 10 mA.

It can be seen that the foward voltage drop across a diode is higher than for an ordinary signal diode; in addition the reverse breakdown voltage is smaller, typically in the range 3 to 5 V.

An LED is very efficient and it does not get hot while it is lit. It is not affected by shock or vibration and does not have a surge current when turned ON. For these reasons LEDs are long-life devices.

Most LEDs are employed in a digital circuit and operated from a 5 V, or increasingly a 3 V, power supply. It will often prove necessary to connect a resistor in series with the LED to limit the current flowing through it to a safe value. The low reverse breakdown voltage means that if an LED is operated from an a.c. supply it will be necessary to protect it with a reverse-connected diode, as shown by Fig. 2.20.

Example 2.8

An LED has the following parameters: $I_{F(max)}$ = 35 mA, $I_{F(typ)}$ = 20 mA, $V_{R(max)}$ = 5 V, $P_{D(max)}$ = 120 mW, and a V_F of 2 V when I_F = 20 mA. The device is to be operated from a 5 V supply. Calculate (*a*) the required value of series resistance, (*b*) the power dissipated in the LED.

Solution

(*a*) $R = (5-2)/(20 \times 10^{-3}) = 150\ \Omega$ (*Ans.*)
(*b*) $P = 2 \times 20 \times 10^{-3} = 40$ mW   (*Ans.*)

(6) Photo-diodes

A photo-diode is made in much the same way as a silicon-signal diode but its case is given a transparent area through which incident light is able to pass and fall on the p-n junction. Energy is then released in the form of extra hole-electron pair generation. The generated holes and electrons move away from the p-n junction under the influence of the reverse-bias voltage. A current therefore flows that adds to the reverse saturation current. This means that the effect of incident light on a photo-diode is to increase the reverse current, and Figs. 2.21(*a*) and (*b*) show the difference between the reverse-current characteristics of a signal diode and a photo-diode. The symbol for a photo-diode is shown in Fig. 2.21(*c*).

A typical family of reverse-current-voltage characteristics is shown in Fig. 2.22. The curve marked as 'dark current' represents the current

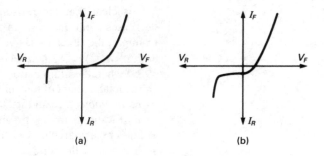

Fig. 2.21 Showing the difference between the reverse-current characteristics of (*a*) a signal diode and (*b*) a photo-diode; (*c*) photo-diode symbol

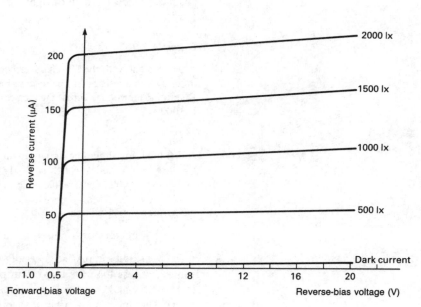

Fig. 2.22 Reverse-current-voltage characteristics of a photo-diode

that flows when there is *no* incident light, i.e. the reverse saturation current of the diode.

The greater the incident luminous flux, the greater is the increase in the reverse current as shown by the curve for any value of lumination, say 1000 lx. If the illumination is kept at a constant level the current increases only slightly as the reverse-bias voltage is increased. If the reverse-bias voltage is reduced to zero and then has its polarity changed to become a forward bias, the current that flows does not alter much until the forward voltage reaches about 0.3 V. If the forward-bias voltage is increased above this value the diode's reverse current will rapidly fall and become zero at about 0.5 V.

It is clear that the resistance of the reverse-biased photo-diode decreases with increase in the incident illumination. When, for example, the reverse-bias voltage is 8 V, the resistance varies from $8/(2.5 \times 10^{-6}) = 3.2$ MΩ when the diode is dark to $8/(205 \times 10^{-6}) = 39$ kΩ when the illumination is 2000 lux.

The total amount of light emitted, or received, by a surface is known as its *luminous flux* and its strength is measured in lumens (lm), or in milliwatts [1 lm = 1.5 mW]. (One lumen is the luminous flux emitted by a source of 1 candela within a solid angle of 1 steradian. A steradian is the solid angle subtended by the centre of a sphere of radius r by an area πr^2 on the surface of the sphere. A complete sphere contains 4π steradians.)

The amount of luminous flux incident upon a surface is measured in lux (lx) or in mW/m^2 [1 lx = 1 mW/m^2]. 1 lx is the illumination produced when 1 lumen of light flux is incident on a surface of area 1 m^2.

Example 2.9

A photo-diode with the I/V characteristics shown in Fig. 2.21 is connected in series with a resistor R. The reverse-bias voltage is 4 V and R is 20 000 Ω. Determine the voltage across the load resistor when the incident illumination is 1000 lx.

Solution

A d.c. load line should be drawn on the characteristics between the points $I = 0$, $V = 4$ V and $V = 0$, $I = 4/20\ 000 = 200$ μA. The load line cuts the curve for 1000 1x where the diode current is 100 μA and the diode voltage is 2 V.

Therefore, load voltage $= 100 \times 10^{-6} \times 2 \times 10^4 = 2$ V (*Ans.*)

The parameters of a photo-diode are

(*a*) Sensitivity in nA/lx; generally quoted for a reverse-bias voltage somewhat less than the maximum permitted value. A typical value is 11 nA/lx at $V_R = 15$ V for $V_{R(max)} = 18$ V.

(*b*) The maximum permitted reverse-bias voltage and reverse current; typically these are 20 V and 14 mA respectively.

(*c*) The maximum forward current; typically 10 mA.

(*d*) The spectral response.

Any photo-electric device is sensitive to incident light, or it radiates light, only in a certain range of wavelengths. Its sensitivity is normally expressed in the form of a graph of output plotted to a base of wavelength. Such a graph is often known as the *spectral response* of the device and Fig. 2.23 gives a typical example. It can be seen that the photo-diode has its maximum sensitivity at wavelengths of about 0.53 μm to 0.66 μm or to green, yellow, orange and red light. In the case of an LED the graph would generally be referred to as its *emission spectrum* graph.

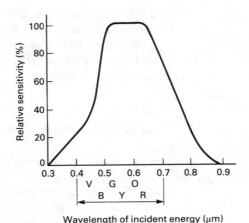

Fig. 2.23 Spectral response of a photo-diode

(7) PIN Diodes

The PIN diode consists of two heavily doped regions of silicon, one n-type and the other p-type, that are separated from one another by a region of (very nearly) intrinsic silicon (see Fig. 2.24(a)). The construction of a PIN diode is shown by Figs. 2.24(b) and (c). The features of the device are a very low forward resistance, a very high reverse resistance, a high reverse breakdown voltage (generally in excess of 100 V and perhaps as high as 1000 V), and a low reverse capacitance. The capacitance is almost constant with change in the reverse-bias voltage, typically changing only about 0.01 pF around a 0.2 pF mean value, for a voltage change of about 100 V.

Because of its low capacitance the main application of the PIN diode is as an electronic switch at frequencies as high as several hundred megahertz and upwards. In the forward direction the resistance of

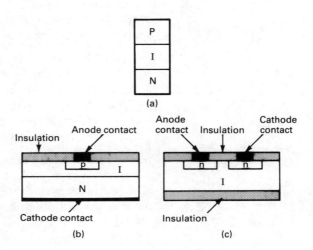

Fig. 2.24 Construction of a PIN diode

a PIN diode might typically vary over the range 1 Ω to 10 kΩ as the forward current is varied from 10 mA to 1 μA. When operated only with a forward bias PIN diodes find application as current-controlled resistors and are employed, for example, in radio equipment in circuits known as attenuators.

(8) Schottky Diodes

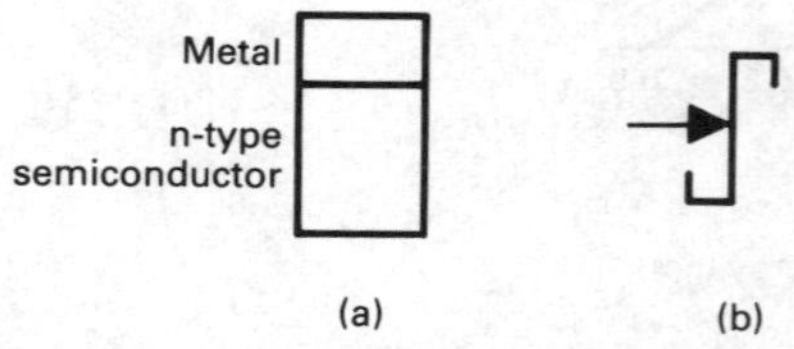

Metal

n-type
semiconductor

(a)

(b)

Fig. 2.25 (a) Schottky diode and (b) symbol for a Schottky diode

A Schottky diode is made by forming a metal, often platinum, region to an n-type silicon region as shown by Fig. 2.25(a). The symbol for a Schottky diode is given in Fig. 2.25(b). Such a device has almost zero charge storage and this means that its reverse recovery time is extremely small. The diode is hence suitable for use in high-speed switching applications where fairly large forward currents are involved. When the diode is forward biased, electrons move from the n-type silicon region into the metal region.

The current-voltage characteristic of a Schottky diode is similar to that for a power diode except that the threshold voltage is approximately 0.3 V. The higher the reverse breakdown voltage of a Schottky diode, the higher will be the forward voltage drop. Typical figures are: (i) 50 V breakdown voltage, 0.5 V forward voltage (as opposed to about 0.8 V for a junction power diode); (ii) 15 V and 0.3 V.

Data Sheets

Data sheets which can be used as an aid to the selection of the correct diode for a particular application are provided by diode manufacturers. Most diode data sheets are headed by the type number and a descriptive title. The maximum voltage and current ratings, generally at an ambient temperature of 25°C, are given. These figures should not be exceeded, otherwise the diode will probably be damaged. The main parameters that are given in a data sheet are as follows:

$I_{F(AV)}$ The maximum forward current that can flow continuously

I_{FRM} The maximum repetitive peak forward current

V_{RRM} The peak reverse repetitive voltage

V_{BR} The breakdown voltage

t_{rr} The reverse recovery time.

The 1N914 is a silicon planar switching diode. Its data sheet includes the figures given in Table 2.5.

Component distributor's sheets generally give the data in a different form that makes it easier to compare different types of diode. This is shown by Table 2.6.

Table 2.5

Absolute maximum ratings at 25°C

V_{RRM}	75 V	$I_{F(AV)}$	75 mA
I_{FRM}	225 mA	P	250 mW

Maximum electrical characteristics at 25°C

V_{BR} (at 100 μA)	100 V	I_R at V_{RRM}	5 μA
I_R at -20 V	25 nA	t_{rr}	4 ns

Table 2.6

Signal diodes

Type no.	Peak inverse voltage (V)	$I_{F(AV)}$ (mA)	Max. I_R	Application
OA200	50	80	100 nA at 50 V	General
1N914	100	75	25 nA at 20 V	Fast switch
OA90	30	10	1.1 mA at 30 V	High frequencies

Rectifier diodes

Type no.	Peak inverse voltage (V)	$I_{F(AV)}$ (A)	Max. V_F drop	Max. I_R
BY127	650	1	1.1 V at 1 A	10 μA at 650 V
1N4001	50	1	1.1 V at 1 A	10 μA at 50 V
1N4007	1000	1	1.1 V at 1 A	10 μA at 1000 V
1N5406	600	3	1.1 V at 3 A	10 μA at 600 V

Zener diodes

Type no.	Nominal Zener voltage (V)	Tolerance	Power dissipation (mW)
1N3996	5.1	5%	10 W
BZY88	2.7,3,3.3,3.6, 4.3,4.7,5.1,6.2, 6.8,7.5,8.2,9.1, 10,11,12,13,15,16, 18,20,22,24,27,30.	5%	400 mW

Clipping And Clamping Circuits

A signal diode can be used to clip or clamp an input alternating waveform. The term clipping means that the top and/or the bottom of a waveform has been removed when the waveform appears at the output of the circuit. Clamping means that either the positive, or the negative, peak of a repetitive alternating voltage is clamped to some particular value, which may be equal to zero volts. This means that a d.c. voltage is effectively added or subtracted to/from the input waveform.

Clipping

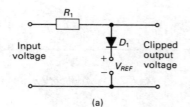

(a)

The basic circuit of a clipper is shown by Fig. 2.26(a). Whenever the input voltage is more positive than the reference voltage V_{REF} the diode D_1 turns ON. The current that then flows through the diode drops the excess input voltage across resistor R_1 with the result that the output voltage has a maximum value equal to V_{REF}. If the input voltage is less positive than V_{REF} or it is of negative polarity, diode D_1 is non-conducting and the input voltage waveform appears at the output terminals of the circuit. Fig. 2.26(b) shows the waveform of the clipped output waveform when the input voltage is of sinusoidal waveform.

Both the positive and the negative half-cycles of the input waveform can be clipped if two diodes are used connected as shown by Fig. 2.27. The two reference voltages need not be of equal magnitude.

(b)

Fig. 2.26 (a) Basic clipping circuit; (b) action of a clipper

Clamping

Two basic clamping circuits are shown in Figs. 2.28(a) and (b). The 'time constant' C_1R_1 of the circuit must be longer than the periodic time of the input waveform. In Fig. 2.28(a) the diode is non-conducting when the input voltage is positive. When the input voltage is negative the diode turns ON and the output voltage is then equal to the small voltage drop V_F across the diode. The capacitor is charged up by the current flowing through the diode and a voltage is developed across it. When the next positive half-cycle arrives the voltage applied across the diode is equal to the sum of the applied input voltage and the voltage across the capacitor minus the diode voltage. After about three cycles of the input waveform the capacitor voltage is equal to the peak value of the input voltage. The average value of the output waveform has shifted in the positive direction so that the least positive value is clamped to just below 0 V. The most positive part of the output voltage is then equal to twice the peak input voltage minus the diode voltage drop, i.e. $2 V - V_F$ volts. If the diode connection is reversed as in Fig. 2.28(b) the positive peak of the input voltage is clamped to very nearly 0 V. Figures 2.28(c) and (d) show the output voltage waveforms for a rectangular input voltage applied to each of the two circuits.

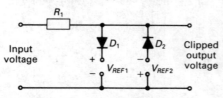

Fig. 2.27 Double polarity clipping circuit

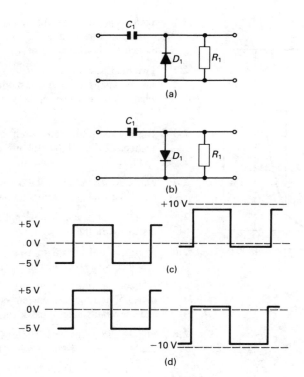

Fig. 2.28 (*a*) and (*b*) basic clamping circuits; (*c*) operation of (*a*); (*d*) operation of (*b*)

If a reference voltage source V_{REF} is connected in series with the diode, an input voltage can be clamped to that voltage instead of to 0 V. Figure 2.29(*a*) shows a clamping circuit that includes a reference voltage V_{REF}. If the input voltage becomes more positive than the reference voltage V_{REF} the diode conducts and this prevents the output voltage from rising any further. The excess voltage is dropped across the capacitor C_1. When the input voltage is at $-V$ volts the positive output terminal of the circuit is at $-V-(V-V_{REF}) = V_{REF} - 2V$ volts. The clamped output voltage is shown by Fig. 2.29(*b*).

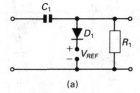

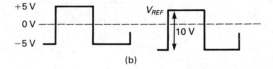

Fig. 2.29 Output voltage may be clamped to a reference voltage

Exercises

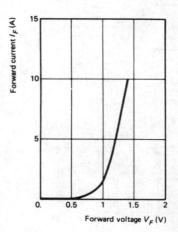

Fig. 2.30

2.1 Sketch a p-n junction and indicate the polarity of the barrier potential. What is its approximate value for a silicon diode? Label the anode and the cathode. Sketch the construction of a planar diode.

2.2 Two diodes, one a germanium type and the other a silicon type, have equal current ratings. If the silicon diode needs 0.70 V to conduct a current of 8 mA, what voltage must be applied to the germanium diode to get the same current flow?

2.3 The reverse saturation current of a diode is 50 nA. Is this a germanium or a silicon device? If the peak inverse voltage is 100 V estimate the breakdown voltage.

2.4 Fig. 2.30 shows a typical diode characteristic. Determine (a) its d.c. forward resistance and (b) its a.c. forward resistance when $V = 1.2$ V.

2.5 A d.c. voltage of 1.1 V is applied to the diode of **2.4**. Calculate the power dissipated in the diode.

2.6 A sinusoidal signal of peak value 1 V is applied to the ideal diode characteristic given in Fig. 2.31. Plot the waveform of the output current.

2.7 A diode has a maximum forward current of 10 A, a forward voltage drop of 1.3 V at a forward current of 3 A, a reverse saturation current of 1 mA and a reverse-voltage breakdown of 300 V. Plot the characteristic of the diode.

2.8 For the diode characteristic of Fig. 2.32 calculate the ratio of the forward d.c. and a.c. resistances when the applied voltage is 0.8 V.

2.9 A Zener diode has a breakdown voltage of 9.1 V and a maximum power dissipation of 1.3 W. Calculate the maximum current the diode should pass.

2.10 Calculate the capacitance of the varactor diode whose characteristic is shown in Fig. 2.33 when the bias voltage is 2 V.

2.11 Fig. 2.32 shows the I/V characteristic of a diode. Determine the change in the diode current when the forward voltage changes from 0.7 to 0.8 V. Hence calculate the a.c. resistance of the diode. A 0.1 V peak sinusoidal voltage is applied to the diode. Plot the current waveform.

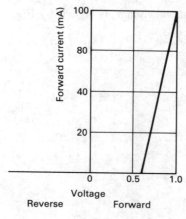

Fig. 2.31

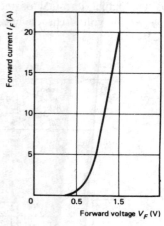

Fig. 2.32

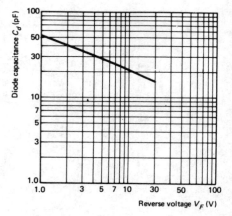

Fig. 2.33

2.12 Each of the diodes shown in Fig. 2.34 is ideal. Calculate the battery current.

2.13 Explain why a diode that is suitable for use as a rectifier of 50 Hz mains supplies is not suitable for use as a switching diode in a logic circuit.

2.14 A semiconductor diode has the data given in Table 2.7.

Table 2.7

Forward voltage (V)	0	0.2	0.4	0.5	0.6	0.7	0.8	0.9
Forward current (mA)	0	0	0.02	0.2	1	8	20	60

The reverse saturation current is 20 nA and the maximum reverse voltage is 50 V. Plot the static characteristic of the diode. Is this a silicon or a germanium diode? Estimate the breakdown voltage of the diode. Suggest a use for the diode.

2.15 A semiconductor diode has the data given in Table 2.8.

Table 2.8

Forward voltage (V)	0	0.4	0.8	1.0	1.2	1.4	1.6	1.8	2.0
Forward current (A)	0	0.03	0.06	0.25	1	6	14	28	70

At the maximum reverse voltage of 300 V the reverse saturation current is 1 μA. Plot the static characteristic of the diode and determine (i) its d.c. forward resistance when the forward voltage is 1 V, and (ii) the a.c. resistance at the point $V = 1.4$ V. What kind of diode is this?

2.16 A Zener diode has the reverse-voltage characteristic given by the data in Table 2.9.

Table 2.9

Reverse voltage (V)	−1	−3	−5	−5.6	−5.7	−5.8	−5.9	−6.0
Reverse current (mA)	0.01	0.01	0.02	1	13	25	37.5	50

Plot the reverse characteristic of the diode. What is the breakdown voltage of the diode? Determine the a.c. resistance of the diode in its breakdown region.

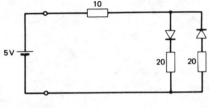

Fig. 2.34

3 Bipolar Transistors

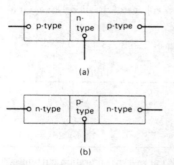

Fig. 3.1 (a) a p-n-p transistor and (b) an n-p-n transistor

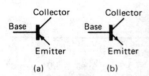

Fig. 3.2 Symbols for (a) a p-n-p transistor and (b) an n-p-n transistor

The transistor is a semiconductor device that can either amplify an electrical signal or act as an electronic switch. Basically a transistor consists of a germanium or silicon crystal which contains three separate regions. The three regions may consist of either two p-type regions separated by an n-type region (Fig. 3.1(a)) or two n-type regions separated by a p-type region (Fig. 3.1(b)). The first type of transistor is known as a p-n-p transistor and the second type as an n-p-n transistor. Both types of transistor are employed, sometimes together in the same circuit, but the discussion throughout this chapter will be in terms of the n-p-n transistor since it is the more commonly employed. However, for the corresponding operation of a p-n-p transistor it is merely necessary to read electron for hole, hole for electron, negative for positive, and positive for negative.

The middle of the three regions in a transistor is known as the *base* and the two outer regions are known as the *emitter* and the *collector*. In most transistors the collector region is made physically larger than the emitter region because it will be expected to dissipate a greater power. The symbol for a p-n-p transistor is given in Fig. 3.2(a) and the symbol for an n-p-n transistor in Fig. 3.2(b). Note that the emitter lead arrowhead is pointing in different directions in the two figures, pointing inwards for the p-n-p transistor and outwards for the n-p-n transistor. The arrowhead indicates the direction in which holes travel in the emitter.

Both p-n-p and n-p-n transistors are generally classified into one of the following groups:

(a) small-signal low-frequency
(b) low-power and medium-power low-frequency
(c) high-power low-frequency
(d) small-signal high-frequency
(e) medium- and high-power high-frequency
(f) switching.

The majority of the transistors listed in manufacturers' and distributors' catalogues are p-n-p silicon types.

The Action of a Transistor

An n-p-n transistor contains two p-n junctions and is normally operated so that one junction, the emmitter-base junction, is forward-biased and the other, the collector-base junction, is reverse-biased. This is shown in Fig. 3.3 together with the directions of the various charge carriers and currents in the transistor. The usual convention whereby the direction of current flow is opposite to the direction of electron movement has been employed.*

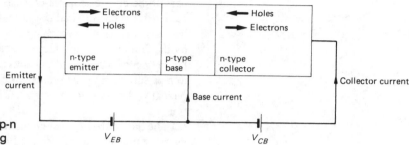

Fig. 3.3 Bias voltages for an n-p-n transistor and the currents flowing

Consider that, initially, the emitter-base bias voltage V_{EB} is zero. Then the majority charge carrier current crossing the emitter-base junction is equal to the minority charge carrier current that is flowing in the opposite direction and the net junction current is zero. The collector-base junction is reverse-biased by the bias voltage V_{CB} and so a small minority charge carrier current flows in the collector lead. This current is the reverse saturation current discussed in the previous chapter but now it is known as the *collector leakage current* and is given the symbol I_{CBO}.

If the emitter-base bias voltage is increased in the negative direction by a few tenths of a volt, the emitter-base junction will be forward-biased and then a majority charge carrier current flows. This current consists of electrons travelling from the emitter to the base and holes passing from the base to the emitter. Only the electron current is useful to the action of the transistor, as will soon be evident, and it is therefore made much larger than the hole current by doping the emitter more heavily than the base. The ratio of the electron current to the total emitter current is known as the *emitter injection ratio* or the emitter efficiency, symbol γ. Typically, γ is approximately equal to 0.995 and this means that only 0.5% of the emitter current consists of holes passing from the base to the emitter.

Immediately the electrons cross the emitter-base junction, and are said to have been emitted or injected into the base, they become minority charge carriers and start to diffuse across the base towards the collector-base junction. Because the base is fairly narrow and is also lightly doped, most of the emitted electrons reach the collector-

---

* d.c. values of current and voltage are indicated by CAPITAL subscripts; a.c. values by lower-case subscripts.

base junction and do not recombine with a free hole on the way. On reaching the junction, the emitted electrons augment the minority charge carrier current crossing the junction and cause an increase in the collector current. The ratio of the number of electrons arriving at the collector to the number of emitted electrons is known as the *base transmission factor*, symbol β. Typically $\beta = 0.995$.

(1) The collector current is less than the emitter current because (*a*) part of the emitter current consists of holes that do not contribute to the collector current and (*b*) not all of the electrons injected into the base are successful in reaching the collector. Factor (*a*) is represented by the emitter injection ratio and factor (*b*) by the base transmission factor; hence the ratio of collector current to emitter current is equal to $\beta\gamma$. Substituting the typical values quoted for γ and β shows that, typically, the collector current is about 0.99 times the emitter current.

(2) The base current is small and has three components: (*a*) a current entering the base to replace the holes lost by recombination with the diffusing electrons, (*b*) the majority charge carrier hole current flowing from base to emitter and (*c*) the collector leakage current I_{CBO}. The first two of these components are currents that flow into the base and together are greater than the leakage current I_{CBO} which flows out of the base, and so the total base current flows into the base. The total current flowing into the transistor must be equal to the total current flowing out of it and hence the emitter current I_E is equal to the sum of the collector and base currents, I_C and I_B respectively, that is

$$I_E = I_C + I_B \tag{3.1}$$

Typically, I_C is equal to $0.99\ I_E$ so that I_B is equal to $0.01\ I_E$.

(3) If the emitter current is varied by some means, the number of electrons arriving at the collector, and hence the collector current, will vary accordingly. The magnitude of the collector-base voltage V_{CB} has relatively little effect on the collector current as will be seen shortly. Control of the output (collector) current can thus be obtained by means of the input (emitter) current and this, in turn, can be controlled by variation of the bias voltage applied to the emitter-base junction. An increase in the forward-bias voltage lowers the height of the potential barrier and allows an increased emitter current to flow; conversely, a decrease in the forward-bias voltage reduces the emitter current.

(4) The ratio of the output current of a transistor to its input current in the absence of an a.c. signal is known as the *d.c. current gain* of the transistor. The output current is the collector current I_C and the input current is the emitter current I_E. Thus,

$$\text{d.c. current gain, } -h_{FB} = I_C/I_E \qquad (3.2)$$

The minus sign indicates that the input and output currents are flowing in opposite directions. By convention, a current flowing into a transistor is taken to be positive and a current flowing out is taken to be negative. Since the operation of the transistor depends on the movement of both holes and electrons, the device is known as a bipolar transistor.

(5) A transistor may be connected in a circuit in one of three ways and in each case one of its three terminals is common to both input and output. The connection is then described in terms of the common terminal; for example, the common-emitter connection has the emitter common to both input and output, the input signal is fed between the base and the emitter, and the output signal is developed between the collector and the emitter. In all connections, the base-emitter junction is always forward-biased and the collector-base junction is always reverse-biased.

The Common-base Connection

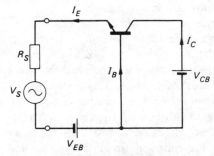

Fig. 3.4 The common-base connection

The basic arrangement of the common-base connection (or configuration) is shown in Fig. 3.4. The transistor has an alternating source of e.m.f. V_S volts r.m.s. and internal resistance R_S ohms connected to its input terminals. The alternating source is connected in series with the emitter-base voltage V_{EB} and it varies the forward bias applied to the emitter-base junction.

During negative half-cycles of the source e.m.f., the forward bias applied to the junction is increased, the potential barrier is lowered and an enhanced emitter current flows into the transistor. Conversely, during positive half-cycles the emitter current is decreased and in this way the collector current is caused to vary in accordance with the waveform of the applied signal voltage. The collector-base bias battery V_{CB} has negligible internal resistance and so the collector-base voltage remains constant as the collector current varies. The collector circuit is said to be short-circuited so far as alternating currents are concerned.

In a common-base amplifier circuit an important parameter is the *short-circuit current gain* of the transistor, symbol h_{fb}. The short-circuit current gain is defined as the ratio of a *change* in collector current to the *change* in emitter current producing it, with the collector-base voltage maintained constant, that is

$$h_{fb} = \delta I_C/\delta I_E = I_c/I_e \text{ when } V_{CB} \text{ is constant} \qquad (3.3)$$

The *short-circuit* current gain is specified since the current gain is a function of the value of any resistance connected in the collector circuit. For the common-base circuit, however, the difference between the short-circuit current gain and the current gain for any particular collector load resistance is very small for all resistance values used in practical circuits and it is usually neglected.

Example 3.1

In a certain transistor a change in emitter current of 1 mA produces a change in collector current of 0.99 mA. Determine the current gain of the transistor.

Solution

$$\text{Current gain } h_{fb} = \delta I_C/\delta I_E = I_c/I_e = 0.99/1 = 0.99 \quad (Ans.)$$

This is a typical value for the short-circuit current gain of a transistor connected in the common-base configuration. It should be evident that h_{fb} must be less than unity, because the emitter current is the sum of the base and collector currents. Clearly, then, a common-base connected transistor must have a current gain of less than unity but, if a resistor is connected in the collector circuit, as shown in Fig. 3.5, both voltage and power gains are possible.

The output voltage is developed across the collector load resistor and, since the internal resistance of the collector supply is negligible, the top end of the resistor is effectively at earth potential so far as alternating currents are concerned. Thus the output signal voltage is taken from between the collector terminal and earth.

The source of the output power is the collector-base bias battery, the transistor effectively acting as a device for the conversion of d.c. power from the battery into the a.c. power supplied to the load.

In the common-base amplifier the input signal voltage and the output signal voltage are in phase with each other, as shown by the waveforms of Fig. 3.5. Consider the input signal voltage to be passing through zero and increasing in the negative direction. The forward bias of the base-emitter junction is then increased and this results in an increase in the emitter current. The collector current is increased and the voltage drop across the collector load resistor R_L increases also and this makes the collector-base potential less positive. Thus a negative increment in the input signal voltage produces a negative increment in the output signal voltage.

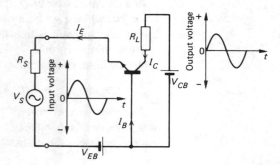

Fig. 3.5 The basic common-base amplifier

The common-base connected transistor is rarely, if ever, used at audio frequencies because of its low current gain and its inconvenient values of input and output impedance.

The Common-emitter Connection

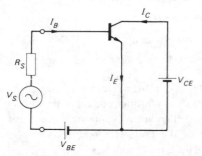

Fig. 3.6 The common-emitter connection

In practice, transistors are most often used in the common-emitter configuration shown in Fig. 3.6.

The emitter-base junction is forward-biased by the battery V_{BE} and the collector-base junction is reverse-biased by a potential equal to $(V_{CE} - V_{BE})$. However, since the voltage of the collector-emitter bias battery V_{CE} is usually much larger than the emitter-base bias voltage V_{BE}, the reverse-bias voltage may often be taken as merely equal to V_{CE} volts.

When a transistor is connected in the common-emitter configuration, the input current is the base current. The operation of the transistor is unchanged from that previously described but the d.c. current gain h_{FE} is the ratio

$$h_{FE} = I_C/I_B \qquad (3.4)$$

During the positive half-cycles of the input signal voltage V_S, the forward bias of the emitter-base junction is increased, and so the emitter current I_E is increased by an amount δI_E. The collector current is also increased, by an amount $\delta I_C = h_{fb}\delta I_E$, and so is the base (input) current, by an amount

$$\delta I_B = \delta I_E - \delta I_C = \delta I_E(1 - h_{fb})$$

Conversely, during negative half-cycles of the input signal voltage all the three currents are reduced in magnitude.

The *short-circuit current gain* of a common-emitter connected transistor, symbol h_{fe}, is defined as the ratio of a *change* in collector current δI_C to the *change* in base current δI_B producing it, the collector-emitter voltage being maintained constant, that is

$$h_{fe} = \delta I_C/\delta I_B = I_c/I_b \text{ when } V_{CE} \text{ is constant} \qquad (3.5)$$
$$h_{fe} = h_{fb}I_e/(I_e - h_{fb}I_e)$$
$$= h_{fb}I_e/I_e(1 - h_{fb})$$
$$= h_{fb}/(1 - h_{fb}) \qquad (3.6)$$

Typical values for the short-circuit current gain h_{fb} of a common-base transistor are in the neighbourhood of unity and thus the common-emitter connection can give a considerable current gain.

The a.c. and the d.c. current gains of a transistor are not usually of the same value, their difference being dependent on the d.c. collector current flowing. Usually, little error is introduced by assuming that h_{FE} is equal to h_{fe}. In any case, the value of h_{FE} (or h_{fe}) can vary considerably between two transistors of the same type and hence of the same nominal current gain.

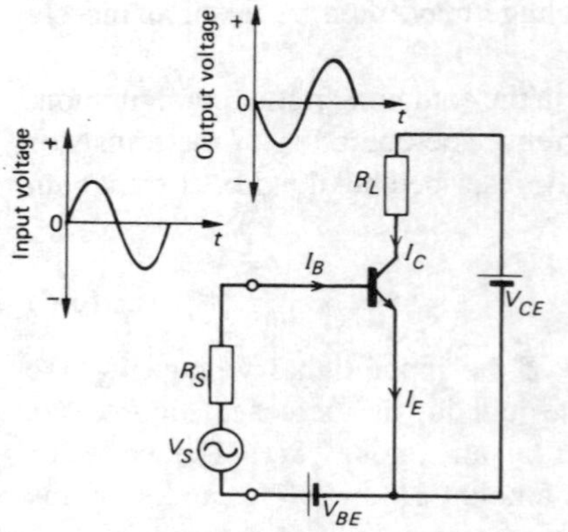

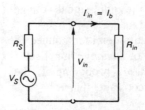

Fig. 3.7 The basic common-emitter amplifier

Fig. 3.8 Circuit for the calculation of the input current to a transistor

Example 3.2

A transistor exhibits a change of 0.995 mA in its collector current for a change of 1 mA in its emitter current. Calculate (*a*) its common-base short-circuit current gain and (*b*) its common-emitter short-circuit current gain.

Solution

(*a*) Common-base short-circuit current gain

$$h_{fb} = I_c/I_e = 0.995/1 = 0.995 \quad (Ans.)$$

(*b*) Common-emitter short-circuit current gain

$$h_{fe} = I_c/I_b = h_{fb}/(1-h_{fb}) = 0.995/(1-0.995) = 199 \quad (Ans.)$$

When a load resistor R_L is connected in the collector circuit (Fig. 3.7) the current gain of the transistor is no longer equal to the short-circuit value but is somewhat less. The actual value of the current gain is dependent on the value of the collector load resistor R_L, decreasing with increase in R_L. The circuit now has both a voltage gain and a power gain.

Fig. 3.8 shows a voltage source of e.m.f. V_S volts (r.m.s.) and internal resistance R_S ohms connected to the input terminals of a transistor whose input resistance is R_{in}.

The a.c. input current to the transistor, I_b, is

$$I_b = V_S/(R_S + R_{in})$$

and the voltage V_{in} appearing across the transistor input terminals is

$$V_{in} = I_b R_{in} \quad \text{or} \quad I_b = V_{in}/R_{in}$$

The output or collector current I_c is

$$I_c = h_{fe} V_{in}/R_{in}$$

This current flows through the collector load resistance R_L and develops the output voltage V_{out} across it, therefore

$$V_{out} = h_{fe} V_{in} R_L/R_{in}$$

and the voltage gain A_v is

$$A_v = V_{out}/V_{in} = h_{fe} R_L/R_{in} \tag{3.7}$$

Since the short-circuit current gain h_{fe} is greater than unity and the collector load resistance R_L is usually greater than the input resistance R_{in} of the transistor, the circuit provides a voltage gain.

The power gain of a common-emitter transistor is the ratio of the power delivered to the load to the power delivered to the transistor. The input power P_{in} to the transistor is (see Fig. 3.8)

$$P_{in} = (I_b)^2 R_{in}$$

and the output power P_{out} is

$$P_{out} = (h_{fe} I_b)^2 R_L$$

Therefore the power gain A_p is

$$A_p = P_{out}/P_{in} = (h_{fe})^2 (I_b)^2 R_L/(I_b)^2 R_{in}$$

$$A_p = (h_{fe})^2 R_L/R_{in} = A_v A_i \qquad (3.8)$$

Again, since R_L is greater than R_{in} the circuit provides a power gain.

Example 3.3

A transistor is connected with common-emitter in a circuit with a collector load resistance of 2000 Ω. The short-circuit current gain of the transistor is 100 and its input resistance is 1000 Ω. Calculate the voltage and power gains of the transistor.

Solution
From equation (3.7),

Voltage gain $= h_{fe}R_L/R_{in} = (100 \times 2000)/1000 = 200$ (*Ans.*)

From equation (3.8),

Power gain $= (h_{fe})^2 R_L/R_{in} = 200 \times 100 = 20\,000$ (*Ans.*)

Example 3.4

A source of e.m.f. 50 mV and internal resistance 1000 Ω is applied to the circuit of example 3.3. Calculate (*a*) the input voltage and (*b*) the output voltage of the circuit.

Solution
(*a*) $V_{in} = (50 \times 1000)/(1000 + 1000) = 25$ mV   (*Ans.*)
(*b*) $V_{out} = A_v V_{in} = 200 \times 25 \times 10^{-3} = 5$ V   (*Ans.*)

The Common-collector Connection

The third way in which a transistor may be connected is shown in Fig. 3.9. The collector terminal is now common to both input and output circuits and the load resistor is connected in the emitter circuit. With this configuration the base current is the input current and the emitter current is the output current. The short-circuit current gain is defined as

Short-circuit current gain

$$h_{fc} = \delta I_E/\delta I_C = I_c/I_b \text{ when } V_{CE} \text{ is constant} \qquad (3.9)$$
$$= I_e/(I_e - I_c)$$
$$= I_e/I_e(1 - h_{fb})$$
$$= 1/(1 - h_{fb}) \qquad (3.10)$$
$$= h_{fe} + 1 \qquad (3.11)$$

But since h_{fe} is a variable it is usual to take h_{fc} as being equal to h_{fe}.

The short-circuit current gain of a transistor connected in the common-collector configuration is approximately equal to the short-circuit gain of the same transistor connected with common-emitter. The current gain when a load is connected in the emitter circuit is not equal to the short-circuit current gain but is reduced by an amount that is dependent on the value of the emitter load.

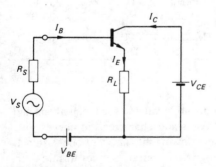

Fig. 3.9 The basic common-collector amplifier

Expressions (3.7) and (3.8) may be used to determine the voltage and power gains of a common-collector circuit.

The input resistance of the common-collector circuit is equal to $h_{fe}R_L + R_{in}$ and so the voltage gain is

$$A_v = A_i R_L/R_{in} = h_{fe}R_L/(h_{fe}R_L + R_{in}) \qquad (3.12)$$

Example 3.5

A transistor is connected in a common-collector circuit with an emitter load resistance of 2000 Ω. The current gain h_{fe} of the transistor is 100 and its input resistance is 1000 Ω. Calculate (a) the voltage gain and (b) the power gain of the circuit.

Solution

(a) $A_v = (100 \times 2000)/(100 \times 2000 + 1000) = 0.995$ (Ans.)

(b) $A_p = 100 \times 0.995 = 99.5$ (Ans.)

The main use of the common-collector circuit — or the *emitter-follower* as it is usually called — is as a buffer which is connected between a high impedance source and a low impedance load.

A comparison between the main parameters of the three transistor configurations is given in Table 3.1.

Table 3.1

Parameter	Common-base	Common-emitter	Common-collector
Short-circuit current gain	h_{fb}	$h_{fe} = h_{fb}/(1-h_{fb})$	$h_{fc} = 1/(1-h_{fb})$
Voltage gain	Good	Better than common-base	Unity or less
Input resistance	Low 30 to 100 Ω	Medium 800 to 5000 Ω	High 5000 to 500 000 Ω
Output resistance	High 10^5 to 10^6 Ω	Medium 10 000 to 50 000 Ω	Low 50 to 1000 Ω

Transistor Static Characteristics

A number of current-voltage plots are available for the analysis and design of a transistor circuit. The static characteristic curves give information on the value of current flowing into or out of one terminal, for either a given current flowing into or out of another terminal, or a given voltage applied between two terminals. Four sets of characteristics can be plotted for each configuration: (a) the input characteristic, (b) the transfer characteristic, (c) the output characteristic and (d) the mutual characteristic. In this book, however, the

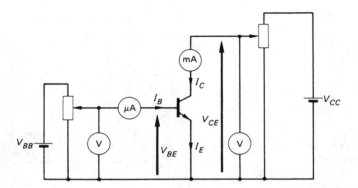

Fig. 3.10 Circuit for the measurement of the static characteristics of a common-emitter connected transistor

characteristics for the common-base and the common-collector circuits will not be discussed. Two versions of set (c) are available.

The method of determining the static characteristics of a transistor is to connect the transistor into a suitable circuit and then to vary the appropriate currents and/or voltages in a number of discrete steps, noting the corresponding values of the other currents at each step. Fig. 3.10 shows a suitable circuit for the measurement of the static characteristics of an n-p-n transistor in the common-emitter configuration.

The collector and base currents are shown as flowing into the transistor and are therefore, by definition, positive; the emitter current is shown flowing out of the transistor and must be taken as negative. If the static characteristics of a p-n-p transistor were to be measured, the polarities of the two batteries would need to be reversed.

(*a*) Common-emitter Input Characteristic

The *input characteristic* shows the way in which the base current varies with change in the base-emitter voltage, the collector-emitter voltage remaining constant. The input characteristic is measured by maintaining the collector-emitter voltage constant at a convenient value and increasing the base-emitter voltage in a number of discrete steps, noting the base current at each step. This procedure is then repeated for a different but constant value of collector-emitter voltage V_{CE}, since any change in this voltage has an effect on the input characteristic. A typical input characteristic is shown in Fig. 3.11. The input resistance, h_{ie}, of the transistor for a given base-emitter voltage V_{BE} is given by the reciprocal of the slope of the curve at that point. For example, consider the input resistance of the transistor at the point $V_{BE} = 0.7$ V.

$$h_{ie} = \delta V_{BE}/\delta I_B = V_{be}/I_b \ (V_{CE} \text{ constant}) \qquad (3.13)$$
$$= 0.02/(22 \times 10^{-6}) = 909 \ \Omega$$

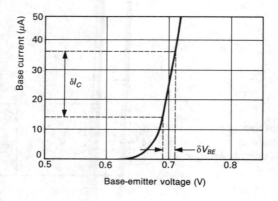

Fig. 3.11 Common-emitter input characteristic
$\delta V_{BE} = 0.71 - 0.69 = 0.02$ V
$\delta I_B = 36 - 14 = 22$ μA

The corresponding p-n-p characteristics would have negative values of I_B, V_{BE} and V_{CE}.

Since the input characteristic is non-linear, the a.c. input resistance of a transistor will vary with the base current. For example,

$$\text{when} \quad I_B = 2 \text{ } \mu\text{A} \quad h_{ie} = 6.7 \text{ k}\Omega$$

Some manufacturers' data sheets give a graph that shows how h_{ie} varies with the collector current.

The d.c. input resistance h_{IE} is given by the ratio V_{BE}/I_B. At the point of measurement

$$h_{IE} = 0.7/(25 \times 10^{-6}) = 28 \text{ k}\Omega$$

It is evident that a considerable difference may exist between the values of h_{IE} and h_{ie}. If the point of measurement is reduced below about $V_{BE} = 0.68$ V, the slope of the input characteristic is much smaller and hence the a.c. input resistance h_{ie} will be larger so that any difference between h_{IE} and h_{ie} will be much less. For base currents of the order of a few tens of microamps, h_{ie} is typically about 3 kΩ.

(b) Common-emitter Transfer Characteristic

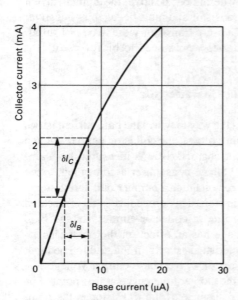

Fig. 3.12 Common-emitter current-transfer characteristic
$\delta I_C = 2.1 - 1.1 = 1$ mA
$\delta I_B = 8 - 4 = 4$ μA

The *transfer characteristic* shows how the collector current changes with changes in the base current, the collector-emitter voltage being held at a constant value. For this measurement the collector-emitter voltage is kept constant and the base current is increased in a number of discrete steps and at each step the collector current is noted. Finally a plot is made of collector current against base current. Since the transfer characteristic is not independent of the value of the collector-emitter voltage, the procedure can be repeated for a number of different collector-emitter voltages to give a family of curves; Fig. 3.12 shows a typical n-p-n transistor transfer characteristic.

The slope of the transfer characteristic gives the short-circuit current gain h_{fe} of the transistor.

$$h_{fe} = \delta I_C/\delta I_B = I_c/I_b = (1 \times 10^{-3})/(4 \times 10^{-6}) = 250$$

(c) Common-emitter Output Characteristic

The *output characteristic* illustrates the changes that occur in collector current with changes in collector-emitter voltage, for constant values of base current. Alternatively, the collector current can be plotted against collector-emitter voltage for constant values of base-emitter voltage. The base current, or the base-emitter voltage, is set to a convenient value and is then maintained constant and the collector-emitter voltage is increased from zero in a number of discrete steps, the collector current being noted at each step. The collector-emitter voltage is then restored to zero and the base current, or the base-emitter voltage, is increased to another convenient value and the procedure repeated. In this way a family of curves (see Figs. 3.13 and 3.14) can be obtained. For the corresponding p-n-p characteristic the polarities of I_C, I_B and V_{CE} should be changed to negative.

The slope $\delta I_c/\delta V_{BE}$ of the output characteristic is the *output admittance*, h_{oe}, of the transistor.

The *output resistance* of the transistor is equal to the *reciprocal* of the slope of the output characteristic. From Fig. 3.13 the output resistance at the point $V_{CE} = 6$ V and $I_B = 30$ μA is

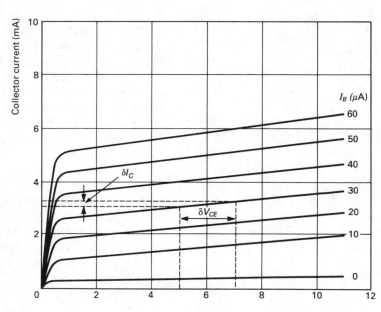

Fig. 3.13 Common-emitter output characteristics
$\delta V_{CE} = 7 - 5 = 2$ V
$\delta I_C = 3.3 - 3.1 = 0.2$ mA

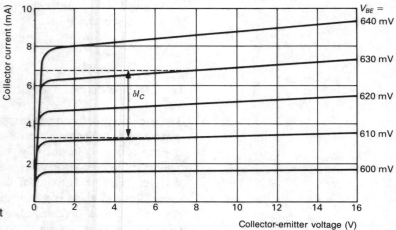

Fig. 3.14 Common-emitter output
characteristics with V_{BE} as input

$$R_{out} = 1/h_{oe} = \delta V_{CE}/\delta I_C = V_{ce}/I_c \ (I_B \text{ constant}) \qquad (3.14)$$
$$= 2/(0.2 \times 10^{-3}) = 10\ 000\ \Omega$$

The output admittance h_{oe}, and hence the output resistance, varies somewhat with the d.c. collector current. Manufacturers' data sheets may include a graph showing h_{oe} plotted against I_C.

When a characteristic is non-linear its slope will vary according to the point of measurement and therefore the point of measurement should always be quoted. It is usual, unless specified otherwise, to measure the slope in the middle of the characteristic. For the greatest accuracy the increments taken either side of the chosen point should be as small as possible although this has not been done in this chapter in order to clarify the diagrams.

The output characteristics of Fig. 3.13 can also be used to determine the short-circuit current gain h_{fe} of the transistor, since, for a given value of collector-emitter voltage V_{CE}, the change in collector current δI_C produced by a change in base current δI_B can be obtained by projecting from the appropriate curves. Thus, for $V_{CE} = 4$ V a change in the base current from 20 μA to 30 μA will produce a change in the collector current from 2.1 to 2.9 mA. The current gain h_{fe} is therefore equal to

$$[(2.9-2.1) \times 10^{-3}]/[(30-20) \times 10^{-6}] = 80$$

The current gain h_{fe} of a transistor is not a constant quantity but varies with the d.c. collector current. Fig. 3.15 shows a typical graph of h_{FE} plotted against I_C. Manufacturer's' data generally quote the collector current at which the maximum value of h_{fe} is obtained, e.g. for the BC 108 the typical h_{fe} is 180 at $I_C = 2$ mA. (Both h_{fe} and h_{FE} vary.)

It will be seen that Fig. 3.13 shows a collector current flowing even when the base current is zero. This current is the *common-emitter*

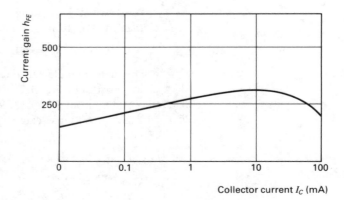

Fig. 3.15 Variation of h_{FE} with d.c. collector current

leakage current, symbol I_{CEO}, which is related to the common-base leakage current I_{CBO} according to the expression

$$I_{CEO} = I_{CBO}(1+h_{FE}) \qquad (3.15)$$

Example 3.6

The collector leakage current I_{CBO} of a transistor is 10 nA. If the d.c. current gain of the transistor is $h_{FB} = 0.995$, calculate (*a*) the collector leakage current when the transistor is connected with common emitter, and (*b*) the d.c. collector current if a d.c. base current of 10 μA is supplied to the base.

Solution

$$h_{FE} = 0.995/(1-0.995) = 199$$
(*a*) $I_{CEO} = 10 \times 10^{-9} \times 200 = 2\ \mu\text{A}$ (*Ans.*)
(*b*) $I_C = h_{FE}I_B + I_{CEO} = 199 \times 10 \times 10^{-6} + 2 \times 10^{-6}$
$$= 1.992\ \text{mA}\quad (\textit{Ans.})$$
(Clearly the contribution of the collector leakage current is negligibly small.)

The collector leakage current I_{CBO} of a common-base transistor is extremely temperature sensitive and is approximately doubled for every 12 °C rise in temperature for silicon transistors and every 8 °C rise for germanium transistors. However, the leakage current of a silicon transistor at a given temperature is much less than the leakage current of an equivalent germanium transistor at the same temperature.

Typically, I_{CBO} at 20 °C may be about 10 μA for a germanium transistor but only about 10 nA for a silicon transistor.

(*d*) Common-emitter Mutual Characteristic

The *mutual characteristics* of a common-emitter connected transistor show the changes in collector current that occur with changes in the base-emitter voltage, with the collector-emitter voltage held constant. Fig. 3.16 shows a typical mutual characteristic. The slope of the mutual characteristic is the mutual conductance g_m of the transistor.

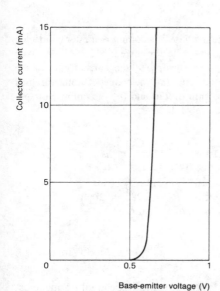

Fig. 3.16 Common-emitter mutual characteristic

Thus, when V_{BE} changes from 0.6 V to 0.65 V the resulting change in I_C is from 1.5 mA to 11.5 mA and so

$$g_m = \delta I_C/\delta V_{BE} = (10 \times 10^{-3})/0.05 = 200 \text{ mS}$$

The mutual conductance can also be determined from the output characteristics of Fig. 3.14 using a similar method to that employed to obtain h_{fe} from Fig. 3.13. When $V_{CE} = 8$ V, a change in V_{BE} from 610 mV to 630 mV causes I_C to vary from 3.3 mA to 6.7 mA. Hence

$$g_m = \delta I_C/\delta V_{BE} = (3.4 \times 10^{-3})/(20 \times 10^{-3}) = 170 \text{ mS}$$

$$\text{Also} \quad g_m = \delta I_C/\delta V_{BE} \tag{3.16}$$
$$= (\delta I_C/\delta I_B) \times (\delta I_B/\delta V_{BE})$$
$$= h_{fe}/h_{ie} \tag{3.17}$$

The mutual conductance g_m depends only on the d.c. collector current I_C, according to the equation

$$g_m = I_C/26 \text{ mS} \tag{3.18}$$

where I_C is in mA.

Thus for the transistor just considered $I_C = (3.3 + 6.7)/2 = 5$ mA, and $g_m = 5/26 = 192$ mS.

From equation (3.7), the voltage gain of a transistor is given by $A_v = A_i R_L/R_{in}$. This can be written as $A_v = h_{fe}R_L/h_{ie}$ and, using equation (3.16)

$$A_v = g_m R_L \tag{3.19}$$

Example 3.7

An 18 mV peak signal is applied to the base of a common-emitter connected transistor. The resulting change in the collector current is 2 mA. Calculate (a) the mutual conductance of the transistor, (b) the current gain of the transistor, (c) the d.c. collector current, (d) the value of collector load resistance that will give a voltage gain of 220 and (e) the output voltage. The transistor has an input resistance of 1200 Ω.

Solution
(a) $g_m = (2 \times 10^{-3})/(18 \times 10^{-3}) = 111$ mS (*Ans.*)
(b) $h_{fe} = g_m h_{ie} = 111 \times 10^{-3} \times 1200 = 133$ (*Ans.*)
(c) $I_C = 26g_m = 26 \times 111 \times 10^{-3} = 2.9$ mA (*Ans.*)
(d) $220 = 111 \times 10^{-3}R_L$, therefore $R_L = 1982 \approx 2000$ Ω (*Ans.*)
(e) $V_{out} = 220 \times 18 \times 10^{-3} = 3.96$ V (*Ans.*)

Thermal and Frequency Effects

Frequency Characteristics

The current gain of a transistor is not the same value at all frequencies, but, instead, falls off at the higher frequencies (see Fig. 3.17). The frequency at which the magnitude of the current gain h_{fe} has fallen to $1/\sqrt{2}$ times its low-frequency value h_{feo} is known as the *cut-off*

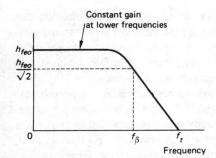

Fig. 3.17 Variation of $|h_{fe}|$ with frequency

frequency f_β of the transistor. Eventually, at some frequency f_t, $|h_{fe}|$ falls to unity. The high-frequency performance of a transistor is often quoted by the manufacturer in terms of a parameter f_t, where

$$f_t = |h_{fe}| \cdot f \qquad (3.20)$$

f being any frequency.

f_t is known as the *transition frequency* or as the *common-emitter gain-bandwidth product*.

Example 3.8

A transistor has $f_t = 500$ MHz. What is its current gain at (*a*) 100 MHz (*b*) 10 MHz?

Solution

  (*a*) $f_t = 500$ MHz $= |h_{fe}| \times 100$ MHz
       $|h_{fe}| = 5$ (*Ans.*)

  (*b*) $f_t = 500$ MHz $= |h_{fe}| \times 10$ MHz
       $|h_{fe}| = 50$ (*Ans.*)

Fig. 3.18 shows how the transition frequency of a typical transistor varies with the collector current. Clearly, if the device is to operate at a high frequency, the collector current must be carefully chosen.

Fig. 3.18 Variation of f_t with collector current

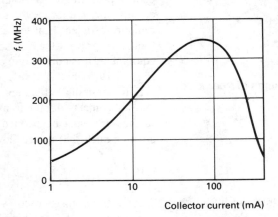

Collector current (mA)

Thermal Runaway

An increase in the temperature of the collector-base junction will cause the collector leakage current I_{CBO} to increase. The increase in collector current produces an increase in the power dissipated at the collector-base junction and this, in turn, further increases the temperature of the junction and so gives a further increase in I_{CBO}. The process is cumulative and, particularly in the common-emitter connection of a germanium transistor, could lead to the eventual destruction of the transistor. In practice, thermal runaway is prevented

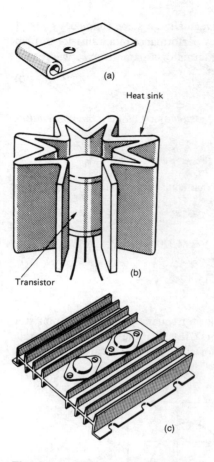

(a)

Heat sink

Transistor

(b)

(c)

Fig. 3.19 Three kinds of heat sink

in well-designed circuits by the use of d.c. stabilization circuitry that compensates for any increase in I_{CBO} and also, for power transistors, by the use of a heat sink to provide rapid conduction of heat away from the junction. Thermal runaway is rarely a problem with silicon transistors because their leakage current is so small.

The manufacturer of a transistor quotes the maximum permissible power that can be dissipated within the transistor without damage. The power dissipated within a transistor is predominantly the power which is dissipated at its collector-base junction, and this, in turn, is equal to the d.c. power taken from the collector supply voltage minus the total output power (d.c. power plus a.c. power). For transistors handling small signals, the power dissipated at the collector is small and generally there is little problem. When the power dissipated by the transistor is large enough to cause the junction temperature to rise to a dangerous level, it is necessary to improve the rate at which heat is removed from the device. Power transistors are constructed with their collector terminal connected to their metallic case. To increase the area from which the heat is removed the case of the transistor can be bolted on to a sheet of metal known as a heat sink. Heat will move from the transistor to the heat sink by conduction and is then removed from the sink by convection and radiation.

The simplest heat sink is shown in Fig. 3.19(a). It consists of a push-fit clip that can be screwed on to a metal chassis, the transistor being a push-fit into the hole. The greater the power dissipated within a transistor, the larger the surface area of the heat sink required to remove sufficient heat to keep the junction temperature within safe limits. To prevent the heat sink occupying too much space within an equipment, it is common to employ structures of the kinds shown in Figs. 3.19(b) and 3.19(c).

For maximum efficiency a heat sink should (i) be in excellent thermal contact with the transistor case, (ii) have the largest possible surface area and be painted matt black and (iii) be mounted in a position such that a free flow of air past it is possible.

The Bipolar Transistor as a Switch

When a transistor is used as a switch it is either biased to be non-conducting or OFF, or it is biased to conduct the maximum possible collector current or be ON. If the base current of a transistor with a collector resistor R_L is gradually increased from zero the collector current will increase as well, since $I_c = h_{FE}I_B$. The collector-emitter voltage V_{CE} will fall since, from Fig. 3.20, $V_{CE} = V_{CC} - I_C R_L$. Eventually, the point will be reached at which V_{CE} has fallen to its minimum possible value. This minimum voltage is known as the collector saturation voltage $V_{CE(SAT)}$. The transistor is then said to be saturated; the collector current at saturation is labelled as $I_{C(SAT)}$ and the base-emitter voltage that *just* produces saturation is labelled as $V_{BE(SAT)}$. Typically, $V_{BE(SAT)}$ is 0.7 V and $V_{CE(SAT)}$ is 0.2 V.

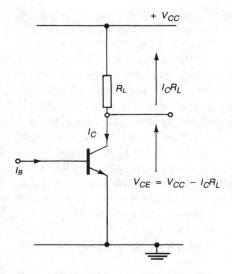

Fig. 3.20 Voltages in a simple bipolar transistor circuit

Typical variations of $V_{CE(SAT)}$ and $V_{BE(SAT)}$ with change in collector current are shown by Figs. 3.21(a) and (b) respectively.

Fig. 3.22 shows a typical set of output characteristics for a transistor. A d.c. load line has been drawn on the characteristics between the points $V_{CE} = V_{CC} = 6$ V, $I_C = 0$ and $I_C = V_{CC}/R_L = 6/1000 = 6$ mA, $V_{CE} = 0$. When a transistor is used as an amplifying device, its operation is restricted to the linear part of its characteristics in order to minimize distortion of the applied signal. When used as a switch, a transistor is rapidly switched between two states. When the base current is zero the transistor is held in its OFF condition. When the transistor is OFF it only conducts the very small collector leakage current. The voltage then dropped across the collector resistor is negligibly small and so the voltage across the transistor in the OFF state is equal to the collector supply voltage V_{CC}. When the transistor is driven into saturation it is then in its ON state. The voltage across the transistor is now its collector *saturation voltage* $V_{CE(SAT)}$.

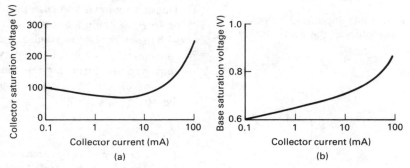

Fig. 3.21 Variation of (a) $V_{CE(SAT)}$ and (b) $V_{BE(SAT)}$ with change in collector current

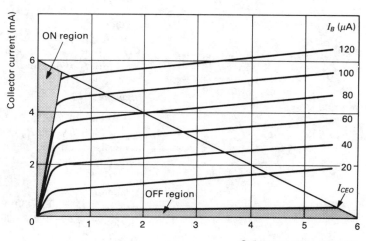

Fig. 3.22 The transistor as a switch

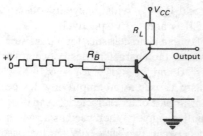

Fig. 3.23 Switching a transistor

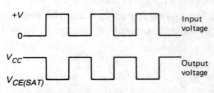

Fig. 3.24 Input and output waveforms of the transistor switch

The transistor can be switched rapidly between its ON and OFF states by the application of a rectangular voltage waveform to its base-emitter terminal (Fig. 3.23). When the input waveform is at zero potential with respect to earth, the transistor will be switched into its OFF state; the voltage which appears at the output terminals is then equal to the collector supply voltage since there will be zero voltage drop across R_L. When the transistor is switched into its ON state by the rectangular base signal voltage, the transistor will conduct heavily and a large voltage, approximately equal to V_{CC}, is dropped across R_L. The output voltage of the circuit is then equal to the saturation voltage $V_{CE(SAT)}$ of the transistor. The voltage at the output terminals of the circuit switches rapidly between $V_{CE(SAT)}$ and V_{CC} volts. It should be noted that when the input signal voltage is positive the output voltage is $V_{CE(SAT)}$, and when the input signal voltage is zero the output voltage is V_{CC} volts (See Fig. 3.24). This means that the input waveform has been inverted; and the circuit has performed the logical function NOT.

The more positive (less negative) voltage level can be regarded as representing logical 1 and the less positive level as giving logical 0. This convention is known as positive logic. Conversely, logical 1 can be taken as being the less positive, or more negative, voltage level with the more positive voltage being labelled as logical 0. This second convention is known as negative logic. Clearly, the convention employed in a particular case must be clearly stated, or understood. Positive logic is the more commonly employed and is employed in the rest of this book.

Example 3.9

The transistor used in the circuit of Fig. 3.23 has $V_{CE(SAT)} = 0.15$ V, $R_L = 1200\ \Omega$ and $V_{CC} = 5$ V. Calculate (a) the saturated collector current $I_{C(SAT)}$ and (b) the power dissipated in R_L when the transistor is (i) ON and (ii) OFF.

Solution

    (a) $I_{C(SAT)} = (5 - 0.15)/1200 = 4.042$ mA (*Ans.*)
    (b) (i) $P = (4.042 \times 10^{-3})^2 \times 1200 = 19.6$ mW (*Ans.*)
         (ii) $I_C = 0$ so $P = 0$ (*Ans.*)

Example 3.10

The circuit of Fig. 3.23 has $V_{CC} = 12$ V, $R_L = 1$ kΩ, $V_{CE(SAT)} = 0.2$ V, $V_{BE(SAT)} = 0.7$ V and $R_B = 30$ kΩ. If a 5 V voltage is applied to the input terminals determine the minimum value of h_{FE} which will produce saturation.

Solution

$$I_B = (5 - 0.7)/(30 \times 10^3) = 143\ \mu\text{A}$$
$$I_{C(SAT)} = (12 - 0.2)/1000 = 11.8\text{ mA}$$

Therefore

$$h_{FE(MIN)} = (11.8 \times 10^{-3})/(143 \times 10^{-6}) = 82.5\quad(\textit{Ans.})$$

Speed of Switching

Rapid switching between the ON and the OFF states of a transistor is desirable in order to minimize the power dissipation within the device. The power dissipated when the transistor is OFF is $P_{OFF} = V_{CC}I_{CEO} \simeq 0$, and when it is ON the dissipation is $P_{ON} = V_{CE(SAT)}I_{C(SAT)}$. Since $V_{CE(SAT)}$ is only a fraction of a volt the power dissipated in either state is very small. Most of the power dissipation occurs while the transistor is actually changing state and so the time taken to achieve this should be as small as possible.

The switching speed of a transistor is the time that elapses between the application of a voltage to the base-emitter terminals and any resulting change in the output voltage. The concept is illustrated by Fig. 3.25. The base-emitter voltage pulse, Fig. 3.25(a), does not cause the collector-emitter voltage, Fig. 3.25(b), to change its state instantaneously. Instead, it remains at the supply voltage V_{CC} volts for a short time and has only fallen to 0.9 V_{CC} after a time period t_d. This is the time needed to fully charge the capacitance of the base-emitter p-n junction. After t_d seconds the collector-emitter voltage V_{CE} falls at a rate that is limited by the need to charge the collector-base capacitance, but eventually it reaches its saturation value $V_{CE(SAT)}$. The *fall-time* t_f is the time taken for V_{CE} to fall from 0.9 V_{CC} to 0.1 V_{CC}. The *turn-on time* t_{ON} is the sum of the delay time and the fall-time, i.e. $t_{ON} = t_d + t_f$.

Once the collector-emitter voltage has fallen to its saturation value any further increase in the base current will not give a corresponding increase in the collector current. The *excess charge* is stored in the base region of the transistor. The base storage effect ensures that when the base-emitter voltage is reduced to zero the transistor does not cease conduction instantaneously. There is an initial *storage delay* t_s during

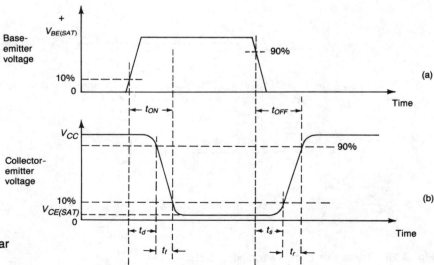

Fig. 3.25 Switching a bipolar transistor

which the excess base charge is removed from the transistor. At the end of this period the collector current falls, and so the collector-emitter voltage rises towards V_{CC} volts. The rate at which the collector-emitter voltage is able to increase is limited by the need for the junction and load capacitances to be discharged. The *risetime* t_r is the time taken for the collector-emitter voltage to increase from $0.1\ V_{CC}$ to $0.9\ V_{CC}$ volts.

The *turn-off time* t_{OFF} is equal to the sum of the storage time and the risetime, i.e. $t_{OFF} = t_s + t_r$.

Typical figures for the turn-on and the turn-off times of a transistor are $t_{OFF} = 360$ ns and $t_{ON} = 55$ ns for a general-purpose transistor and $t_{OFF} = 125$ ns and $t_{ON} = 27$ ns for a switching transistor.

Switching Circuits

A simple switching circuit can be designed using a single transistor with the load in the collector circuit and a current-limiting resistor in the base circuit. A basic lamp-control circuit is shown in Figs 3.26(*a*) and (*b*).

When the base current is zero there will be zero collector current and the lamp will not light. When a voltage pulse is applied, Fig. 3.26(*a*), or the switch is closed, Fig. 3.26(*b*), a base current, and hence a collector current, flows and the lamp lights. The transistor that is employed should have adequate current gain, a low collector saturation voltage, and, if high-speed operation is desired, low values of turn-on and turn-off times. The value of the base resistor R_B should be chosen to ensure that the transistor saturates. The base current required will need to be somewhat greater than $I_{C(SAT)}/h_{FE}$ because the value of h_{FE} will fall as the collector-emitter voltage approaches its saturation value. The minimum value of h_{FE} obtained from the data sheet should be used.

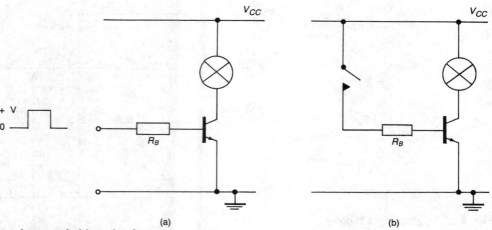

(a)   (b)

Fig. 3.26 Transistor lamp switching circuits

Applying Kirchoff's law to the base circuit of Fig. 3.26(a) gives

$$V = I_B R_B + V_{BE},$$

or

$$R_B = (V - V_{BE})/I_B \qquad (3.21)$$

If I_B is to be the value which just produces the saturation collector current $I_{C(SAT)}$ then V_{BE} should be the saturation value $V_{BE(SAT)}$ obtained from the data sheet.

Example 3.11

In the circuit given in Fig. 3.26(a) the collector supply voltage V_{CC} is 5 V and the resistance of the lamp when lit is 50 Ω. The transistor parameters are $h_{FE} = 100$, $V_{BE(SAT)} = 0.7$ V and $V_{CE(SAT)} = 0.2$ V. Calculate the necessary value of the base resistor if the applied voltage is 5 V.

Solution
The saturated collector current is

$$I_{C(SAT)} = (5 - 0.2)/50 = 96 \text{ mA}.$$

The minimum base current for saturation is

$$I_{C(SAT)}/h_{FE} = (96 \times 10^{-3})/100 = 960 \ \mu\text{A}.$$

From equation (3.21),

$$R_B = (5 - 0.7)/(960 \times 10^{-6}) = 4479 \ \Omega$$

A lower value should be selected to ensure the transistor saturates. The nearest preferred value is 4300 Ω (*Ans.*)

These circuits power the load whenever the transistor has been turned ON by a HIGH input. The circuit shown in Fig. 3.27 passes current through the load when the input is LOW; then T_1 is OFF and the HIGH voltage at its collector turns T_2 ON.

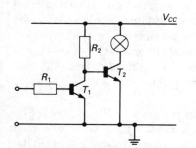

Fig. 3.27 Two-transistor switching circuit

Inductive Loads

When an inductive load, such as an electromagnetic relay coil, is to be switched ON and OFF, precautions must be taken against high voltages that might damage the transistor. When the current flowing through an inductance is suddenly switched off a back e.m.f. $e = -L di/dt$ volts is induced into the inductance. If the rate of change di/dt of the current is high a high voltage may easily be produced. The polarity of this voltage is such that it will try to drive a current through the transistor from the collector to the emitter and this could easily destroy the transistor. To guard against this a diode is always connected across the inductance in the direction shown by Fig. 3.28. When the high voltage back e.m.f. is generated the diode conducts and provides a path for the reverse current, keeping it away from the transistor.

(*Note* the electromagnetic relay is still employed in modern electronic circuitry whenever its perfect isolation properties are required.)

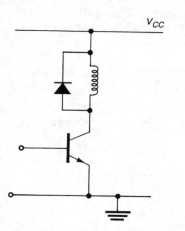

Fig. 3.28 Transistor switching an inductive load

Example 3.12

Draw a lamp circuit that automatically switches ON when it is dark if (*a*) a 6 V, 60 mA (*b*) a 240 V, 60 W lamp is used.

Solution

(*a*) The lamp may be operated directly by a transistor and a possible circuit is given by Fig. 3.29(*a*). When the *light-dependent resistor* (LDR) is in the dark its resistance is high and the base-emitter voltage of T_1 is positive enough to turn T_1 ON. Collector current flows through the lamp and the lamp glows. When it is light the LDR's resistance falls and so the base-emitter voltage of T_1 also falls. The collector current of T_1 falls also and the lamp goes out.

(*b*) It is now necessary to use a relay to operate the 240 V lamp and a possible circuit is shown by Fig. 3.29(*b*).

The Photo-transistor

A *photo-transistor* is fabricated in a similar way to an ordinary bipolar transistor except that its base region may not be brought out to an external terminal (and if it is it is used for bias purposes only), and its packaging has a transparent 'window' that allows external light to be incident on the base region. With the base open-circuited the collector current is the collector leakage current $I_{CEO} = I_{CBO}(1 + h_{FE})$. When the base region is illuminated by light energy extra hole-electron pairs are generated and this leads to an increase in I_{CBO}. In turn, this gives a much greater increase in I_{CEO} and a collector current flows.

The symbol for a photo-transistor is shown by Fig. 3.30(*a*). Fig. 3.30(*b*) gives a set of typical collector current against collector-emitter voltage curves. It can be seen that the characteristics are very similar to the output characteristics of a bipolar transistor, differing in that curves for different values of incident light illumination are shown instead of curves for different base currents. The small current that flows when there is zero incident light is known as the *dark current*.

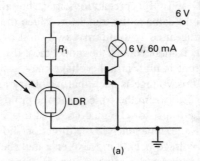

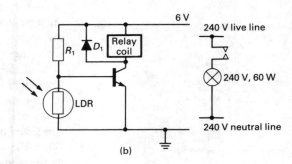

(a)   (b)

Fig. 3.29

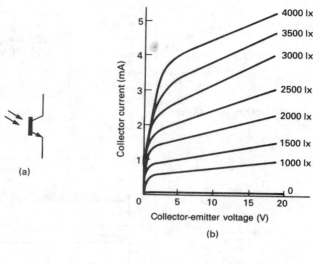

Fig. 3.30 (a) Symbol for a photo-transistor, (b) photo-transistor output characteristics

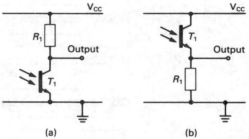

Fig. 3.31 Two photo-transistor circuits

A photo-transistor can be operated in any of the usual transistor configurations, and Figs. 3.31(a) and (b) show two possible circuits.

A photo-transistor is slower to operate than a photo-diode and it can only be used at frequencies up to about 100 kHz. On the other hand, the photo-transistor is more sensitive and it can be directly connected to small loads.

The Construction of Transistors

A number of different methods of manufacturing transistors have been developed since the invention of the transistor in 1948 and a description of many of them is outside the scope of this book. The most commonly employed type of transistor is the silicon planar transistor, and only the construction of this type will be considered.

The construction of a silicon planar transistor is shown in Fig. 3.32. The steps involved in the manufacture of a silicon planar transistor are as follows: a wafer of n-type silicon is oxidized to a depth of approximately 1 micron (Fig. 3.33(a)) and then the oxide is partially etched off (Fig. 3.33(b)). Next the wafer is exposed to a vapour of the acceptor element boron and the impurity is allowed to diffuse into the wafer to a predetermined depth. At the same time the wafer surface

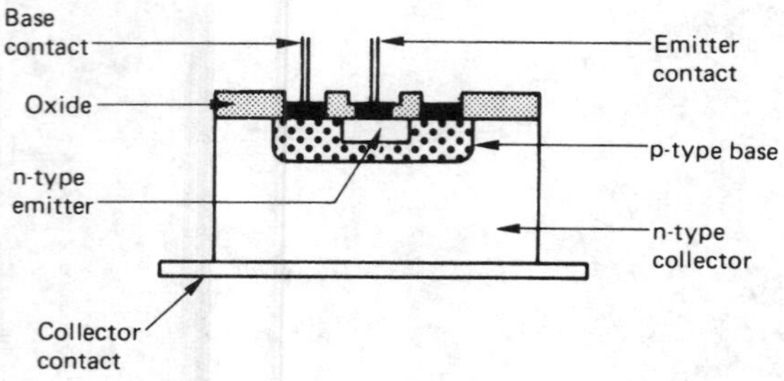

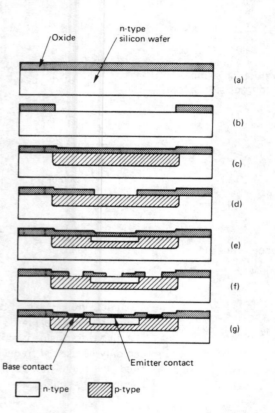

Fig. 3.32 Construction of a silicon planar transistor

Fig. 3.33 The stages in the manufacture of a silicon planar transistor

is reoxidized (Fig. 3.33(c)). A part of the reoxidized surface is then etched away (Fig. 3.33(d)) and the wafer is exposed to a vapour of the donor element phosphorus; it is also reoxidized again (Fig. 3.33(e)). The wafer now contains a layer of p-type material that will form the base of the transistor and a layer of n-type material that provides the emitter. The wafer is then etched to separate the base and emitter regions on the surface of the wafer (Fig. 3.33(f)) and finally (Fig. 3.33(g)) metal contacts are alloyed on to the etched areas.

The wafer is cut to the required size, mounted on a suitable collector contact, and then leads are connected to the base and emitter contacts.

The construction of a planar transistor requires a relatively thick collector wafer in order to give adequate mechanical support to the other layers and this increases the resistance of the collector. For some applications the collector resistance is too large and must be reduced. If the resistance were to be reduced by the use of a low-resistivity material for the collector, the breakdown voltage of the transistor would also be reduced, and the collector-base capacitance would be increased, both undesirable effects. To overcome these effects an epitaxial layer is employed in the collector. This is a layer, approximately 0.1 mm thick, of high-resistivity material that is deposited on the main collector and that permits the collector to be made from a material of low resistivity.

A high-power transistor will be required to conduct a large current and to be able to withstand a large collector-emitter voltage. These requirements introduce certain problems with the basic fabrication method and have led to the introduction of techniques which involve the use of two, or more, epitaxial layers.

Data Sheets

Information about the ratings and the characteristics of a transistor are given by the manufacturer in the form of a data sheet. The data is given by quoting typical figures for the various parameters.

A number of symbols are used in data sheets with a system of subscripts that indicate the transistor terminals to which each symbol refers. The first symbol denotes the terminal at which the current or voltage is measured. The second subscript indicates which of the other two terminals is the reference terminal. A third subscript, when present, is O, S or R; it indicates that the third terminal is, respectively, open-circuit, short-circuit or has a specified value of resistance connected between the third terminal and the reference terminal. Thus, V_{CEO} indicates the maximum collector-emitter voltage with the base open-circuited, V_{CBO} is the maximum collect-base voltage with the base open-circuit, V_{CES} indicates the maximum collector-emitter voltage with the base short-circuited to the emitter.

$I_{B(max)}$, $I_{C(max)}$, and $I_{E(max)}$ are the maximum permissible values of the base, collector and emitter currents.

Some of the data that is given in a data sheet refers to parameters and characteristics that are beyond the scope of this book, e.g. data on noise performance and on thermal characteristics. These are omitted from the simplified data sheet for the BC 147 (Table 3.2) on page 66. A data sheet has a heading that states the number of the device and which type of transistor it is. The main intended application for the transistor is also stated. A number of characteristics then follow. These may include (i) the transfer characteristic, (ii) h_{FE} plotted to base of I_C, (iii) f_t plotted to base of I_C, (iv) $V_{CE(SAT)}$ plotted against I_C, (v) h_{ie} plotted against I_C, (vi) h_{fe} plotted against I_C and (vii) h_{oe} plotted against I_C.

Table 3.2

BC 147 Silicon Planar Epitaxial Transistor
Quick Reference Data

$V_{CES(max)}$	50 V	$V_{CEO(max)}$	45 V	$I_{CM(max)}$	200 mA
$P_{tot(max)}(T_{amb} \leq 25°C)$	350 mW	$T_{J(max)}$	125°C	h_{fe}	125–500
f_t	300 MHz				

Ratings

$V_{CBO(max)}$	50 V	$V_{CES(max)}$	50 V	$V_{CEO(max)}$	45 V
$V_{EBO(max)}$	6 V	$I_{C(max)}$	100 mA	$I_{CM(max)}$	200 mA
$I_{EM(max)}$	200 mA	$I_{BM(max)}$	200 mA	$P_{tot(max)}$ $(T_{amb} \leq 25°C)$	350 mW

Electrical Characteristics
($T = 25°C$ unless otherwise stated)

I_{CBO}	$V_{CB} =$	20 V,	$I_E = 0$		15 nA (max.)	—	—
V_{BE}	$I_C =$	2.0 mA,	$V_{CE} = 5.0$ V	550 mV (min.)	620 mV (typ.)	700 mV (max)
	$I_C =$	10 mA,	$V_{CE} = 5.0$ V	—	—	770 mV (max)
$V_{CE(SAT)}$	$I_C =$	10 mA,	$I_B = 0.5$ mA	—	90 mV (typ.)	250 mV (max)
	$I_C =$	100 mA,	$I_B = 5.0$ mA	—	200 mV (typ.)	600 mV (max)
$V_{BE(SAT)}$	$I_C =$	10 mA,	$I_B = 0.5$ mA	—	700 mV (typ.)	
	$I_C =$	100 mA,	$I_B = 5.0$ mA	—	900 mV (typ.)	
h_{FE}	$I_C =$	10 mA,	$V_{CE} = 5.0$ V	40 (min.)	90 (typ.)	
	$I_C =$	2 mA,	$V_{CE} = 5.0$ V	110 (min.)	180 (typ.)	220 (max)

h parameters ($I_C = 2.0$ mA, $V_{CE} = 5.0$ V, $f = 1000$ Hz)

h_{ie} (input impedance)	1600 Ω (min.)	2700 Ω (typ.)	4500 Ω (max)
h_{fe} (a.c. current gain)	125 (min.)	220 (typ.)	260 (max)
h_{oe} (output admittance)		18×10^{-6} s (typ.)	30×10^{-6} s (max)

Component distributors provide the more important data in a somewhat more compact form which makes it easier for the potential user to compare the parameters of different types of transistor. The data is presented in the manner shown by Table 3.3.

Selection of a Transistor

The main points to be considered in the choice of a particular type of transistor are its intended application, that is, whether the device is to be operated as an amplifier or as a switch, the frequency of operation, and the currents, voltages and powers that it will have to be able to handle.

In general, a small-signal low-frequency transistor will work satisfactorily as either an amplifier or a switch provided that the collector current is small. The same transistor would not, however, be able to dissipate high power and would not have a low saturation voltage when the collector current is high.

(1) If a transistor is required for a small-signal audio-frequency

Table 3.3

A. *Small-signal low-frequency n-p-n transistors*

Type no.	V_{CEO} (V) (max)	V_{CBO} (V) (max)	V_{EBO} (V) (max)	I_C (mA) (max)	$V_{CE(SAT)}$ (V)	P_{tot} (mW) (max)	Typ. h_{FE} at I_C	Typ f_t (MHz)
BC107	45	50	6	100	0.25	300	290 at 2 mA	300
BC108	20	30	5	100	0.25	300	520 at 2 mA	300
BC147	45	50	6	200	0.25	220	180 at 2 MA	300

B. *Medium-power low-frequency n-p-n transistors*

BC142	60	80	5	800	1.0	800	20 at 200 mA	40
BC337	200	250	5	100	1.0	800	60 at 30 mA	80
BFY51	30	60	6	1000	0.35	800	40 at 150 mA	50

C. *High-power low-frequency n-p-n transistors*

BD131	45	70	6	3000	—	1500	40 at 500 mA	60
BD135	45	45	5	1000	0.6	8000	100 at 150 mA	250
TIP31A	60	60	5	3000	—	40W	25 at 3 A	3

D. *Small-signal high-frequency n-p-n transistors*

BF115	30	50	5	30	—	145	40 at 1 mA	230
BF180	20	30	3	20	—	150	*	675
ZTX326	12	25	—	50	—	200	20 at 25 mA	1000

E. *Medium- and high-power n-p-n transistors*

BF258	250	250	5	100	—	800	25 at 30 mA	90
2N3866	30	55	4	400	—	5000	100 at 50 mA	—

F. *Medium-current switching n-p-n transistors*

BFX84	60	100	6	1000	1.0	800	110 at 150 mA	50
BSX20	15	40	4	500	0.6	350	80 at 10 mA	600
2N2219	40	75	6	800	1.0	800	200 at 150 mA	250

* Gain is expressed in a different way.
Note: most general-purpose small-signal transistors are also good switches, e.g. the BC108.

amplifier, the only parameters of importance will be the current gain and perhaps the maximum collector-base voltage. This is because the powers involved will be small enough to be handled easily by any transistor and high-frequency operation is not required. In most cases the transistor voltages are unlikely to be an important factor since the voltages in the amplifier will probably be well within the capabilities of most a.f. transistors. f_t is unimportant, although it ought to be greater than the product $h_{fe} \times f_{max}$.

(2) When choosing an audio-frequency power transistor, care is necessary to ensure that the power dissipation expected within the transistor will be well within the manufacturer's quoted maximum value, and it is quite likely that the maximum collector-base voltage will also need careful consideration.

(3) A transistor selected for use in a radio-frequency amplifier should have an f_t that is several times higher than the highest frequency of operation. Then, suitable current, voltage and power ratings can be taken into account.

(4) A transistor chosen for a switching application should have a low collector saturation voltage; usually, a general-purpose device will suffice unless a large current is to be switched. If so, a specific medium-current switching transistor must be employed. If high-speed switching is required the turn-on and turn-off times will need consideration.

Transistor Arrays

A transistor array consists of a number of separate transistors inside a dual-in-line integrated circuit package. One example of such an array, the CA 3083, is shown by Fig. 3.34. It can be seen to contain five separate transistors. Besides being very economical both with costs and with physical space, the use of a transistor array has the advantage that since the transistors are all formed in a single silicon wafer they are thermally closely matched. This means that any changes in their parameters, such as current gain or leakage current, due to a change in temperature, is the same for each transistor.

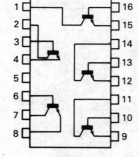

Fig. 3.34 CA 3083 transistor array

Exercises

3.1 An n-p-n transistor has an emitter current of 3.2 mA and a current gain h_{FB} of 0.99. Calculate the base current.

3.2 A p-n-p transistor has an emitter current of 3.2 mA and a base current of 100 μA. Calculate the collector current.

3.3 One of the transistors shown in Fig. 3.35 is conducting. Is it (a), (b), (c) or (d)?

3.4 Explain what is wrong with the following statement. An n-p-n transistor is operated with its base-emitter junction forward biased and its collector-base

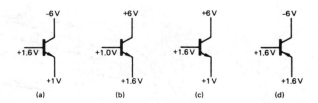

Fig. 3.35

junction reverse biased. When the base potential is made more negative, the collector current is increased in value.

3.5 Explain why the collector current of a transistor is *always* smaller than the emitter current.

3.6 A transistor has an emitter injection ratio of 0.997 and a base transmission factor of 0.996. Calculate its short-circuit current gain in the common-emitter configuration.

3.7 Draw the block diagram of an n-p-n transistor with its base-emitter junction forward biased and its collector-base junction reverse biased. Mark on the diagram the directions of (a) the base current, (b) electrons in the collector region and (c) holes in the emitter region.

3.8 Draw the circuit of a basic common-base amplifier using an n-p-n transistor. Explain how the signal voltage applied to the emitter is amplified.

3.9 The application of a signal voltage of 7.5 mV peak between the base and emitter terminals of an n-p-n transistor causes the emitter current to vary by ± 0.5 mA about its d.c. value. If $h_{fb} = 0.99$ calculate the a.c. voltage developed across a 1200 ohm load resistor connected in the collector circuit. Calculate the voltage gain of the circuit.

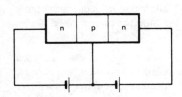

Fig. 3.36

3.10 For the n-p-n transistor shown in Fig. 3.36 indicate the directions of (a) the emitter, base and collector currents, (b) holes in the base region and (c) electrons in the two batteries.

3.11 A transistor has a current gain of 150 and a voltage gain of 300. Calculate whether its power gain is (a) 150, (b) 45 000, (c) 45 or (d) 2.

3.12 A transistor has $h_{fe} = 350$. Calculate h_{fb}.

3.13 A transistor has a base current of 10 μA, a current gain of 90 and a collector leakage current I_{CBO} of 5 nA. Calculate its collector current.

3.14 An n-p-n transistor has $h_{fb} = 0.998$. Calculate its h_{fe} value. What would be the h_{fe} of a p-n-p transistor having the same value of h_{fb}?

3.15 A transistor has $h_{fe} = 250$, and an input resistance h_{ie} of 1200 ohms. It is connected in a circuit with a load resistor of 1000 ohms. Calculate its power gain.

3.16 An n-p-n transistor has an input resistance of 1000 ohms and it is connected to a source of e.m.f. 50 mV and impedance 1000 ohms. Calculate (a) the base current, (b) the signal base-emitter voltage.

3.17 A transistor has a current gain h_{fe} of 400. Calculate its current gain h_{fb}.

3.18 A transistor has $h_{FE} = 120$. Calculate h_{FC}.

3.19 A transistor has $h_{FB} = 0.985$. Calculate both h_{FE} and h_{FC}.

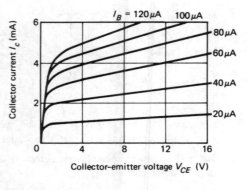

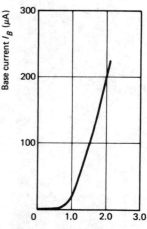

Fig. 3.37

Fig. 3.38 Base-emitter voltage V_{BE} (V)

Fig. 3.39

3.20 Derive from the output characteristics given in Fig. 3.37 the transfer characteristics for the transistor. Is the device n-p-n or p-n-p?

3.21 Fig. 3.38 shows the input characteristic of a transistor. Calculate its input resistance when $V_{BE} = 1$ V, 1.2 V, 1.4 V, 1.6 V and 1.8 V and then plot input resistance against base current.

3.22 A transistor has an output resistance of 15 kilohms and its operating point is $V_{CE} = 6$ V, $I_C = 2$ mA. What will be the collector current when $V_{CE} = 8$ V?

3.23 Draw a typical set of output characteristics for an n-p-n transistor and then use them to explain (a) how the collector current depends on the base current, (b) how the collector current depends on the collector-emitter voltage.

3.24 Fig. 3.39 gives the mutual characteristics of an n-p-n transistor. Determine the minimum, typical and maximum values of the mutual conductance of the transistor when the collector current is 20 mA.

3.25 Data for a transistor is given in Table 3.4. Plot curves showing how (a) the current gain and (b) the mutual conductance vary with the collector current.

3.26 The input and transfer characteristics of a transistor are given in Fig. 3.40. Use the characteristics to find the collector current when the base-emitter voltage is 0.7 V.

3.27 Fig. 3.41 shows the mutual characteristics of a transistor. (a) What type of transistor is it? (b) Calculate the mutual conductance when the base-emitter voltage is (i) 700 mV, (ii) 750 mV.

Table 3.4

Base-emitter voltage (V)	0.5	0.55	0.6	0.65
Base current (μA)	0	1	12	70
Collector current (mA)	0	0.08	1.5	11.5

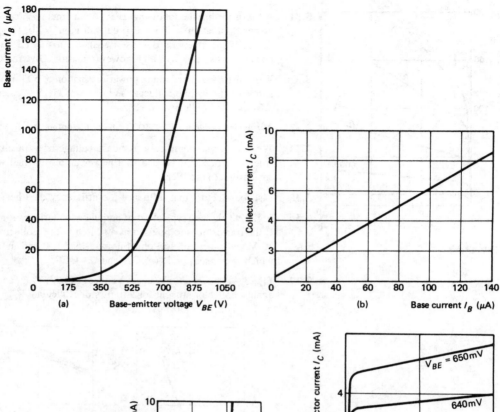

Fig. 3.40

(a) Base-emitter voltage V_{BE} (V)

(b) Base current I_B (μA)

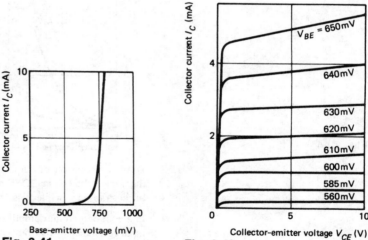

Fig. 3.41

Fig. 3.42

3.28 For the output characteristics shown in Fig. 3.42 calculate the output resistance of the transistor when the collector-emitter voltage is 5 V and the base-emitter voltage is (i) 620 mV and (ii) 640 mV.

3.29 For the transistor whose characteristics are given in Fig. 3.41 calculate the mutual conductance when the collector-emitter voltage is 5 V and the base-emitter voltage is 620 mV.

3.30 For the transistor referred to in Fig. 3.42 determine the value of the collector current for each value of the base-emitter voltage when $V_{CE} = 6$ V. Then plot the mutual characteristic of the transistor when the collector-emitter voltage is 6 V.

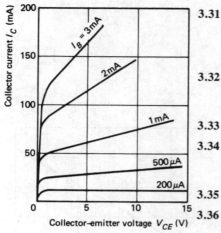

Fig. 3.43

3.31 A transistor has the following data: (i) maximum $V_{CB} = 10$ V, (ii) maximum power dissipation = 1 W, (iii) current gain = 40, (iv) $f_t = 20$ MHz. The device is (a) a general-purpose transistor, (b) an r.f. amplifier, (c) a low-power transistor or (d) a high-power transistor. Give reasons for your choice.

3.32 A transistor has a maximum power dissipation of 500 mW. If the d.c. voltage applied between the collector and the base is 15 V calculate the maximum possible d.c. collector current.

3.33 A transistor has an f_t of 1000 MHz. Calculate its h_{fe} at 100 MHz.

3.34 Why is the maximum collector-base voltage of a transistor more important when the transistor is used as a power amplifier than when it is used as a small-signal amplifier?

3.35 List four factors that influence the efficiency of a heat sink.

3.36 Fig. 3.43 shows a typical set of output characteristics for an n-p-n transistor. If the transistor is to be used as a switch with a collector supply voltage of 15 V and a load of 150 ohms draw a load line and indicate the regions in which the transistor is (a) ON and (b) OFF.

3.37 Fig. 3.44 shows a typical set of curves of (a) $V_{CE(sat)}$ and (b) $V_{BE(sat)}$ plotted against collector current for a transistor. Calculate $V_{CE(sat)}$ and $V_{BE(sat)}$ for $I_C = 1$ A and $I_B = 50$ mA.

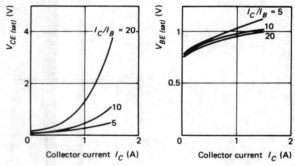

Fig. 3.44

3.38 Draw a typical set of output characteristics for an n-p-n transistor and on them mark the ON and OFF regions. Explain why little power is dissipated in the transistor while it is in either the ON or the OFF state.

3.39 In a common-emitter amplifier the current gain of the transistor is 120, the input resistance of the transistor is 3 kΩ and the collector load resistance is 2 kΩ. Determine the voltage and power gains of the amplifier.

3.40 Figs. 3.45(a) and (b) show, respectively, the transfer and mutual characteristics of a transistor. Determine (a) the current gain h_{fe}, (b) the mutual conductance g_m and (c) the input resistance h_{ie} of the device.

3.41 The data given in Table 3.5 refer to a transistor in the common-emitter configuration.

Use the data to plot the output characteristics for $V_{BE} = 600$ mV, 610 mV and 620 mV. Use the characteristics to determine: (a) the output resistance of the transistor for $V_{BE} = 610$ mV, (b) the mutual conductance for $V_{CE} = 6$ V.

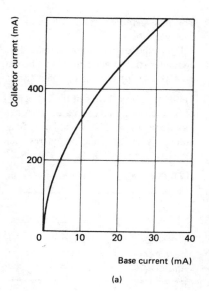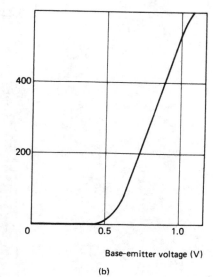

Fig. 3.45

(a)   (b)

Table 3.5

Collector-emitter voltage (V)	Collector current (mA)		
	Base-emitter voltage 600 mV	Base-emitter voltage 610 mV	Base-emitter voltage 620 mV
1	3.1	4.6	6.0
3	3.5	5.1	6.6
5	3.9	5.6	7.2
7	4.3	6.1	7.8
9	4.7	6.6	8.4

Table 3.6

Collector-emitter voltage (V)	Collector current (mA)		
	Base current $-10\,\mu A$	Base current $-20\,\mu A$	Base current $-30\,\mu A$
−2	−2.9	−4.4	−5.9
−4	−3.3	−4.9	−6.4
−6	−3.7	−5.4	−7.0
−8	−4.1	−5.9	−7.6
−10	−4.5	−6.4	−8.2

3.42   The data given in Table 3.6 refer to a transistor in the common-emitter configuration.

Draw the output characteristic for I_B = −10, −20, −30 μA. Use the characteristics to determine: (a) the output resistance of the transistor for I_B = −20 μA, (b) the current gain for V_{CE} = −7 V.

3.43   Table 3.7 gives values of the collector current-collector voltage for a series of base current values in a transistor in the common-emitter configuration. Plot these characteristics and hence find (i) the current gain when the collector voltage is 6 V, (ii) the output resistance for a base current of 45 μA.

Table 3.7

Collector-emitter voltage (V)	Collector current (mA)			
	Base current 25 μA	Base current 45 μA	Base current 65 μA	Base current 85 μA
3	0.91	1.59	2.25	3.00
5	0.92	1.69	2.45	3.20
7	0.96	1.84	2.65	3.50
9	0.99	2.04	2.95	4.00

4 Field-effect Transistors

The field-effect transistor (FET) is a semiconductor device which can perform all of the functions of a bipolar transistor, but which operates in a fundamentally different way. There are four kinds of FET available: the junction field-effect transistor (JFET), the insulated gate field-effect transistor (IGFET) and the vertical metal-oxide-silicon power field-effect transistor (VMOS). The IGFET is more often known as the metal-oxide-silicon field-effect transistor (MOSFET). This latter term will be used throughout this book. The MOSFET can be sub-divided into two classes: the enhancement type and the depletion type. All four classes of FET can be obtained in either n-channel or p-channel versions and so a total of eight different types of FET are available. Two newer devices are the insulated gate bipolar transistor (IGBT) and the metal semiconductor field-effect transistor (MESFET).

The Junction Field-effect Transistor

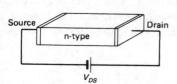

Fig. 4.1 n-type semiconductor

Fig. 4.1 shows a wafer of lightly doped n-type silicon, provided with an ohmic contact at each of its two ends, and a d.c. voltage V_{DS} applied between these contacts. The contact to which the positive terminal of the power supply is connected is known as the *drain*, while the negative side of the supply is connected to the *source*.

A current, consisting of majority charge carriers, will flow in the silicon wafer from drain to source. The magnitude of the drain current is inversely proportional to the resistance of the wafer. In turn, the resistance of the wafer depends on the resistivity of the n-type silicon wafer and the length l and cross-sectional area a of the conduction path, or *channel*, i.e. $R = \rho l/a$. For given values of resistivity and length, the channel resistance is determined by the cross-sectional area of the channel. If, therefore, the cross-sectional area can be varied by some means, the channel resistance and hence the drain current can also be varied.

The properties of a p-n junction are such that the region either side of the junction, known as the *depletion layer*, is a region of high resistivity whose width is a function of the reverse-biased voltage

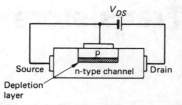

Fig. 4.2 The basic junction FET

applied to the junction. The depletion layer can be used to effect the required control of the channel resistance. A p-n junction is therefore required in the silicon wafer and to obtain one it is necessary to diffuse a p-type region into the wafer, as shown by Fig. 4.2. The p-type region is doped more heavily than the n-type channel to ensure that the depletion layer will lie mainly within the channel. An ohmic contact is provided to the p-type region and it is known as the gate terminal.

If the gate is connected directly to the source, the p-n junction will be reverse-biased and the depletion layer will be extended further into the channel. The p-n junction is reverse-biased because the p-type gate region is at zero potential, while the n-type channel region is at some positive potential. A potential gradient will exist along the length of the channel, varying from a positive value equal to the d.c. voltage supply V_{DS} at the drain end to zero voltage at the source end. Since the cross-sectional area of the channel between the gate region is smaller than at either end of the channel (because of the depletion layer), the resistance of this area is relatively large, and most of the voltage drop appears across this part of the channel. The drain end of the part of the channel which is alongside the gate region is at a higher positive potential than the source end of the channel; hence the reverse-bias voltage applied to the p-n junction is greater on this side. The effect of this voltage on the depletion layer is shown in Fig. 4.3.

When the drain-source voltage is zero (Fig. 4.3(a)), the depletion layer either side of the p-n junction is narrow and it has little effect on the channel resistance. Increasing the drain-source voltage in the positive direction from 0 V will widen the depletion layer and cause it to extend into the channel. This is shown in Fig. 4.3(b), which makes it clear that the layer widens more rapidly at the drain end of the channel than at the source end.

Thus, increasing the drain-source voltage increases the channel resistance. This increase in the channel resistance results in the increase in the drain current being less than proportional to the increase in voltage, i.e. doubling the drain-source voltage V_{DS} does not give

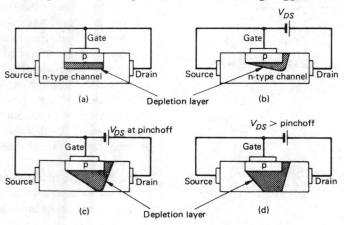

Fig. 4.3 Showing the effect of increasing the drain-source voltage

a two-fold increase in drain current because the channel resistance has increased also. Further increase in the drain-source voltage makes the depletion layer extend further into the channel and eventually the point is reached where the depletion layer extends right across the channel (Fig. 4.3(c)). The drain-source voltage which produces this effect is known as the *pinch-off voltage*. Once pinch-off has developed, further increase in drain-source voltage widens the pinched-off region (Fig. 4.3(d)). The drain current ceases to increase very much with increase in the drain-source voltage, but becomes more or less constant with change in drain-source voltage. The drain current continues to flow because a relatively large electric field is set up across the depleted region of the channel, and this field aids the passage of electrons through the region.

The reverse-bias voltage applied to the gate-channel p-n junction can also be increased by the application of a negative potential, relative to source, to the gate terminal. If the gate-source voltage V_{GS} is made negative, the increase in reverse-bias voltage widens the depletion layer over the width of the gate region and thereby further increases the channel resistance. The drain current therefore falls as the gate-source voltage is made more negative until it is approximately equal to the pinch-off voltage and the channel is *pinched-off* (Fig. 4.4). When this condition occurs the drain current is zero. Generally, the JFET is operated with voltages applied to both its drain and gate terminals, with the drain-source voltage greater than the pinch-off value. The resistance of the channel up to the pinch-off point is determined by the gate-source voltage, and the drain-source voltage produces an electric field which sweeps electrons across the extended depletion layer. The drain current is then more or less independent of the drain-source voltage and is under the control of the gate-source voltage.

A p-channel junction FET operates in a similar manner except that it is necessary to increase the gate-source voltage in the positive direction to reduce the drain current. Also, of course, the drain is held at a negative potential with respect to the source. The symbols used for n-channel and p-channel JFETs are given, respectively, in Figs. 4.5(a) and (b). Both types of JFET are operated with their gate-channel p-n junction reverse-biased; hence they have a very high input impedance.

The JFET as an Amplifier

If a JFET is to operate in an amplifier circuit it must be possible to control its drain current by means of a signal voltage. If the drain current is then passed through a resistance, an output voltage will be developed across the resistance that is an amplified version of the input signal voltage. The necessary control of the drain current can be obtained by connecting the signal voltage in the gate-source circuit of the FET (Fig. 4.6).

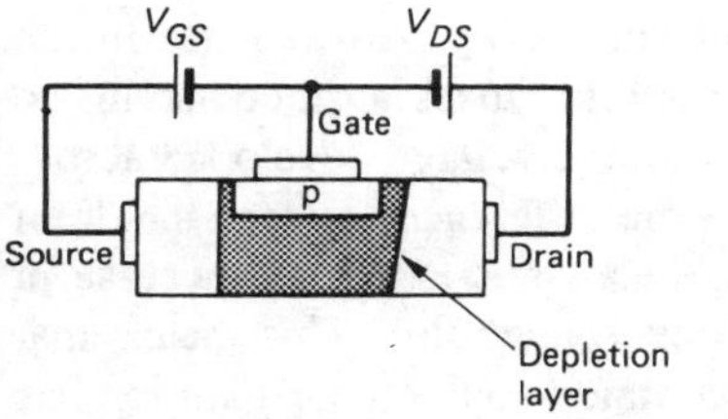

Fig. 4.4 Showing the effect of increasing the gate-source voltage

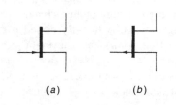

Fig. 4.5 Symbols for (a) an n-channel JFET and (b) a p-channel JFET

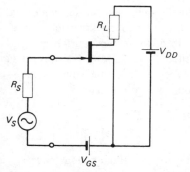

Fig. 4.6 The basic JFET amplifier

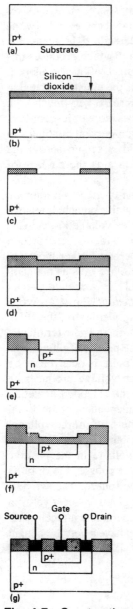

Fig. 4.7 Construction of an n-channel JFET

The signal source, of e.m.f. V_S and impedance R_S, is connected in the gate-source circuit of the FET in series with a d.c. voltage supply V_{GS}. The total reverse-bias voltage applied to the gate-channel junction is the sum of the signal voltage V_S and the bias voltage V_{GS}. During the positive half-cycles of the signal waveform, the reverse-bias voltage applied to the p-n junction is reduced, the depletion layer becomes narrower and so the drain current increases. Conversely, negative half-cycles of the signal waveform increase the bias voltage and cause the depletion layer to extend further into the channel; the drain current is therefore reduced. In this way, the drain current is caused to vary with the same waveform as the input-signal voltage.

The output voltage is developed across the drain load resistor R_L, and can be taken off from between the drain and earth. A voltage gain is achieved because the alternating component of the voltage across R_L is larger than the signal voltage V_S. An increase in the signal voltage in the positive direction produces an increase in the drain current and hence an increase in the voltage developed across R_L. The drain-source voltage V_{DS} is the difference between the drain supply voltage V_{DD} and the voltage across R_L; thus an increase in drain current makes the drain-source voltage fall. This means that a JFET amplifier operated in the common-source configuration has its input and output signal waveforms in antiphase with one another. It is necessary to ensure that the signal voltage is not large enough to take the gate-source voltage positive by more than about 0.5 V, otherwise the high input impedance feature of the JFET will be lost.

The various steps involved in the manufacture of an n-channel JFET are shown by Figs. 4.7(a) through to (g). A heavily doped p-type silicon substrate marked as p^+ in Fig. 4.7(a) has a layer of silicon dioxide grown on to its surface (Fig. 4.7(b)). Next (Fig. 4.7(c)) a part of the silicon dioxide layer is etched away to create an exposed area of the p-type silicon substrate into which n-type impurities can be diffused. An n-type region is thus produced in the p-type substrate and then another layer of silicon dioxide is grown on to the surface (Fig. 4.7(d)). The next steps, shown by Fig. 4.7(e), are first to etch another gap in the silicon dioxide layer and then to diffuse a p^+ region into the exposed area of the n-type region of the substrate. A third layer of silicon dioxide is then grown over the surface of the device (Fig. 4.7(f)). Gaps are now etched into the layer into which aluminium contacts to the two ends of the n-type region and the upper p-type region can be deposited (Fig. 4.7(g)). The terminals connected to the two ends of the n-type region are the source and the drain contacts, while the third terminal acts as the gate.

Parameters

The important parameters of a JFET are its mutual conductance g_m, its input resistance R_{in} and its drain-source resistance r_{ds}. The *mutual conductance* is defined as the ratio of a change in the drain current to the change in the gate-source voltage producing it, with the drain-source voltage maintained constant, i.e.

$$g_m = \delta I_D / \delta V_{GS} = I_d / V_{gs} \quad V_{DS} \text{ constant} \qquad (4.1)$$

The *drain-source resistance* r_{ds} is the ratio of a change in the drain-source voltage to the corresponding change in drain current, with the gate-source voltage held constant, i.e.

$$r_{ds} = \delta V_{DS} / \delta I_D = V_{ds} / I_d \quad V_{GS} \text{ constant} \qquad (4.2)$$

Typically, g_m has a value lying in the range of 1 to 7 mS, while r_{ds} may be 40 kΩ to 1 MΩ. The input impedance of a JFET is the high value presented by the reverse-biased gate-channel p-n junction. Typically, an input impedance in excess of 10^8 Ω may be anticipated. I_{DSS} is the drain current which flows when the gate-source voltage V_{GS} is zero.

Example 4.1

A JFET has a signal voltage of 1.5 V peak value applied to its input terminals. The drain current then varies by ± 2 mA about its quiescent value. Calculate the mutual conductance of the JFET.

Solution
$$g_m = (2 \times 10^{-3})/1.5 = 1.33 \text{ mS} \quad (Ans.)$$
(Note how small this value of g_m is compared with the values obtained for a bipolar transistor).

The Metal-oxide Silicon Field-effect Transistor

The metal-oxide silicon field-effect transistor, generally known as the MOSFET, differs from the JFET in that its gate terminal is insulated from the channel by a layer of silicon dioxide. The layer of silicon dioxide increases the input impedance of the FET to an extremely high value, such as 10^{10} Ω or even more. The high value of input impedance is maintained for all values and polarities of gate-source voltage, since the input impedance does not depend on a reverse-biased p-n junction.

The MOSFET is available in two different forms: the depletion type and the enhancement type. Both types of MOSFET can be obtained in both n-channel and p-channel versions, so that there are altogether four different kinds of MOSFET. Most discrete MOSFETs are of the depletion type and most integrated circuit devices are enhancement-mode types.

Depletion-type MOSFET

The constructional details of an n-channel depletion mode MOSFET are shown in Fig. 4.8. Two heavily doped n^+ regions are diffused into a lightly doped p-type substrate and are joined by a relatively lightly doped n-type channel.

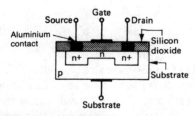

Fig. 4.8 Construction of an n-channel depletion-type MOSFET

The gate terminal is an aluminium plate that is insulated from the channel by a layer of silicon dioxide. A connection is also made via another aluminium plate to the substrate itself. In most MOSFETS the substrate terminal is internally connected to the source terminal but sometimes an external substrate connection is made available. The substrate must always be held at a negative potential relative to the drain to ensure that the channel-substrate p-n junction is held in the reverse-biased condition. This requirement can be satisfied by connecting the substrate to the source. A depletion layer will extend some way into the channel, to a degree that depends on the magnitude of the drain-source voltage. Because of the voltage dropped across the channel resistance by the drain current, the depletion layer extends further across the part of the channel region nearest to the drain than across the part nearest the source. The resistance of the channel depends on the depth to which the depletion layer penetrates into the channel. With zero voltage applied to the gate terminal the drain current will, at first, increase with increase in the drain-source voltage, but once the depletion layer has extended right across the drain end of the channel the drain current becomes, more or less, constant with further increase in the drain-source voltage.

The channel resistance, and hence the drain current, of a depletion-type MOSFET can also be controlled by the voltage applied to the gate. A positive voltage applied to the gate will attract electrons into the channel from the heavily doped n^+ regions at either end. The number of free electrons available for conduction in the channel is increased and so the channel resistance is reduced. The reduction in channel resistance will, of course, allow a larger drain current to flow when a given voltage is maintained between the drain and source terminals. An increase in the positive gate voltage will increase the drain current which flows when the drain-source voltage is large enough to extend the depletion layer across the drain end of the channel. Conversely, if the gate is held at a negative potential relative to the source, electrons are repelled out of the channel into the n^+ regions. This reduces the number of free electrons which are available for conduction in the channel region and so the channel resistance is increased. The drain current that flows when the depletion layer has closed the channel depends on the channel resistance.

The drain current of a depletion-type MOSFET can therefore be controlled by the voltage applied between its gate and source terminals.

Enhancement-type MOSFET

Figure 4.9 shows the construction of an enhancement-type MOSFET. The gate terminal is insulated from the channel by a layer of silicon dioxide, and the substrate and source terminals are generally connected together to maintain the channel-substrate p-n junction in the reverse-biased condition. It can be seen that a channel does not exist between

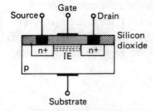

Fig. 4.9 Construction of an n-channel enhancement-type MOSFET (IE = Induced electrons forming a virtual channel when gate voltage is positive)

the n^+ source and drain regions; hence the drain current that flows when the gate-source voltage is zero is very small. If, however, a voltage is applied between the gate and source terminals, which makes the gate positive with respect to the source, a *virtual channel* will be formed. The positive gate voltage attracts electrons into the region beneath the gate to produce an n-type channel (shown dotted in the figure) in which a drain current is able to flow. The positive voltage that must be applied to the gate to produce the virtual channel is called the *threshold voltage* and is typically about 2 V. Once the virtual channel has been formed, the drain current which flows depends on the magnitude of both the gate-source and drain-source voltages. An increase in the gate-source voltage above the threshold value will attract more electrons into the channel region, and a depletion region is formed, Fig. 4.10(*a*). As the drain-source voltage is increased the depletion region is widened until *pinch-off* occurs, see Fig. 4.10(*b*). Further increase in the drain-source voltage gives only a small increase in the drain current and the depletion region widens still more, Fig. 4.10(*c*).

For a particular value of gate-source voltage the drain current will increase with increase in drain-source voltage up to onset of pinch-off and thereafter it will remain more or less constant.

The drain current of a MOSFET can be controlled by the voltage V_{GS} applied between its gate and source terminals: increasing V_{GS} produces an increase in the drain current. If the drain current is passed through a suitable resistance, a voltage gain can be provided. The basic arrangement of a MOSFET amplifier is similar to the JFET circuit given in Fig. 4.6 and it operates in a similar manner.

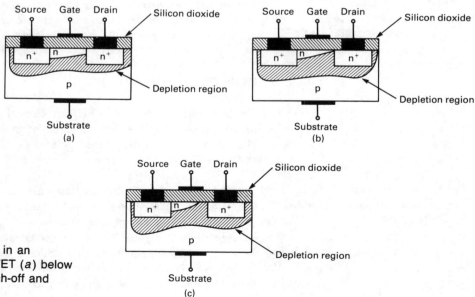

Fig. 4.10 Pinch-off in an enhancement MOSFET (*a*) below pinch-off, (*b*) at pinch-off and (*c*) above pinch-off

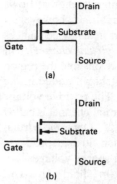

(a)

(b)

Fig. 4.11 Symbols for (a) an n-channel depletion-type MOSFET and (b) an n-channel enhancement-type MOSFET

The important parameters of a MOSFET are the same as those of a JFET: namely, its mutual conductance g_m, its drain-source resistance r_{ds} and its input resistance R_{in}. Typically, g_m is the range $1-10$ mS, r_{ds} is some $5-50$ kΩ and R_{in} is 10^{10} Ω or more. It should be noted that whereas the values of mutual conductance are approximately the same as those of a JFET, the drain-source resistance values are lower but the input resistance is higher.

Figs. 4.11(a) and (b) show the symbols for n-channel depletion-type and enhancement-type MOSFETS. The symbols for the p-channel versions differ only in that the direction of the arrow-head is reversed. (*Note*: in most circuits the substrate of an n-channel MOSFET is connected to the source and the substrate of a p-channel MOSFET is connected to the drain.)

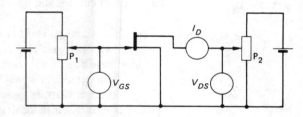

Fig. 4.12 Circuit for the determination of the static characteristics of an n-channel JFET

Static Characteristics

The static characteristics of a FET are plots of drain current against voltage and are used to determine the drain current which flows when a particular combination of gate-source and drain-source voltages is applied. Two sets of static characteristics are generally drawn: these are the drain and the mutual characteristics.

Drain Characteristics

The drain characteristics of a FET are plots of drain current against drain-source voltage for constant values of gate-source voltage. The characteristics can be determined with the aid of a circuit such as that shown in Fig. 4.12 for the measurement of the characteristics of an n-channel junction FET.

The data required to plot the drain characteristics consists of the values of the drain current which flow as the drain-source voltage is increased in a number of discrete steps starting from zero, the gate-source voltage being held constant at a convenient value. The method generally used to obtain the data is as follows: the gate-source voltage is set to a convenient value by means of the potential divider P_1 and then the drain-source voltage is increased, starting from zero, in a number of discrete steps. At each step the drain current flowing is noted. The gate-source voltage is then set to another convenient value and the procedure is repeated. In this way sufficient data can be obtained to plot a family of curves of drain current to a base of the drain-source voltage. This family of curves is known as the drain characteristics of the FET. The drain characteristics of the other types of FET are obtained in a similar manner. Fig. 4.13 shows typical drain characteristics for the six types of FET.

Each curve has a region of small values of V_{DS} in which I_D is proportional to V_{DS}. In these regions the devices can be operated as a voltage-dependent resistance, i.e. as a resistance whose value V_{DS}/I_D depends on the value of V_{DS}. For all devices the drain current which flows when the gate-source voltage is zero is labelled as I_{DSS}.

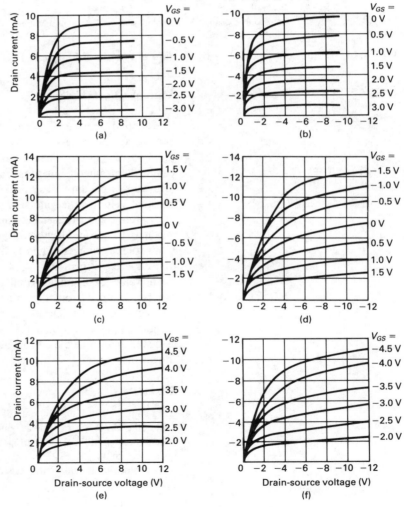

Fig. 4.13 The drain characteristic of
(a) an n-channel JFET,
(b) a p-channel JFET,
(c) an n-channel depletion-type MOSFET,
(d) a p-channel depletion-type MOSFET,
(e) an n-channel enhancement-type MOSFET,
(f) a p-channel enhancement-type MOSFET

[These (a)−(f) labels apply also to Fig. 4.14]

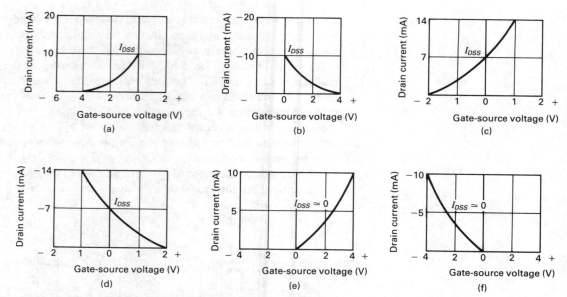

Fig. 4.14 Mutual characteristics of six different kinds of FET [see Fig. 4.13 for types]

Mutual Characteristics

The mutual or transfer characteristics of a FET are plots of drain current against gate-source voltage for various constant values of drain-source voltage. The mutual characteristics of an n-channel JFET can be determined using the arrangement given in Fig. 4.12 and the following procedure. The drain-source voltage is maintained at a constant value as the gate-source voltage is increased in a number of discrete steps. At each step the value of the drain current flowing is noted. The procedure should then be repeated for a number of other drain-source voltages. Typical mutual characteristics for all six types of FET are given in Fig. 4.14.

The values of the mutual conductance and the drain-source resistance can be obtained from the drain characteristics, while the mutual conductance can be determined from the mutual characteristics. The method employed to obtain the values of these parameters is the same as that to determine the current gain and output resistance of a bipolar transistor.

Example 4.2

An n-channel JFET has the data given in Table 4.1.

Plot the drain characteristics and use them to determine the mutual conductance g_m of the device at $V_{DS} = 12$ V. Calculate also the drain-source resistance for $V_{GS} = -2$ V.

Also plot the mutual characteristics and from them obtain g_m at $V_{DS} = 12$ V.

Table 4.1

Drain-source voltage V_{DS} (V)	Drain current (mA)			
	Gate-source voltage $V_{GS} = 0$ V	$= -1$ V	$= -2$ V	$= -3$ V
0	0	0	0	0
4	7	5.0	2.4	0.30
8	10.1	5.9	2.7	0.35
12	10.2	6.2	2.9	0.40
16	10.25	6.3	3.0	0.45
20	10.3	6.35	3.05	0.50
24	10.35	6.4	3.1	0.55

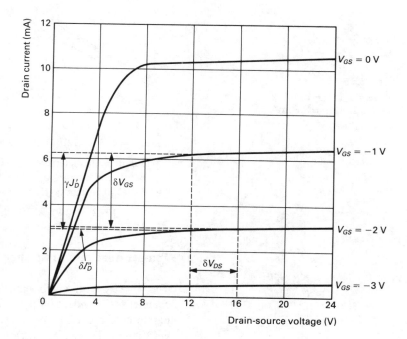

Fig. 4.15
$\delta I'_D = 6.2 - 2.9 = 3.3$ mA
$\delta V_{GS} = -1 - (-2) = 1$ V
$\delta I''_D = 3.0 - 2.9 = 0.1$ mA
$\delta V_{DS} = 8 - 2 = 6$ V

Solution

The drain characteristics of the FET are shown in Fig. 4.15.

The mutual conductance g_m of the FET is given by the expression $g_m = \delta I_D / \delta V_{GS}$, with V_{DS} constant at 12 V. It can be seen from the characteristics that a change in V_{GS} from -2 V to -1 V produces a change in I_D from 2.9 to 6.2 mA. Therefore

$$g_m = [(6.2 - 2.9) \times 10^{-3}]/(2 - 1) = 3.3 \text{ mS} \quad (Ans.)$$

Also from the characteristics it can be seen that a change in V_{DS} from 12 to 16 V, with V_{GS} constant at -2 V, produces a change in I_D from 2.9 to 3.0 mA. Therefore

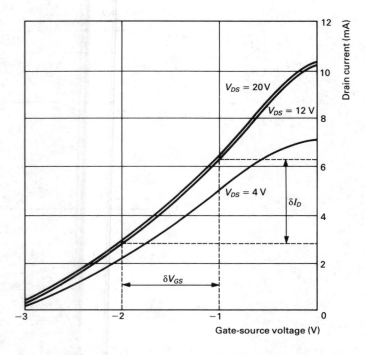

Fig. 4.16
$\delta V_{GS} = -1 - (-2) = 1$ V
$\delta I_D = 6.2 - 2.9 = 3.3$ mA

$$r_{ds} = (16 - 12)/[(3.0 - 2.9) \times 10^{-3}] = 40\ 000\ \Omega \quad (Ans.)$$

The mutual characteristics of the JFET are shown plotted in Fig. 4.16. The mutual conductance g_m of the device is given by the slope of the curve; thus for V_{DS} constant at 12 V,

$$g_m = [(6.2 - 2.9) \times 10^{-3}]/1 = 3.3\ \text{mS} \quad (Ans.)$$

Temperature and Frequency Effects

The velocity with which majority charge carriers travel through the channel is dependent on both the drain-source voltage and the temperature of the FET. An increase in the temperature reduces the velocity of the charge carriers and this appears in the form of a reduction in the drain current which flows for given gate-source and drain-source voltages.

A further factor that may also affect the variation of drain current with change in temperature is the barrier potential across the gate-channel p-n junction. An increase in temperature will cause the barrier potential to fall and this, in turn, will reduce the width of the depletion layer for a given gate-source voltage. The channel resistance will fall and the drain current will increase. The two effects tend to vary the drain current in opposite directions and as a result the overall variation can be quite small. Indeed, it is possible to choose a particular gate-source voltage and obtain zero temperature coefficient. In general, the overall result is that the drain current decreases with increase in temperature. This is the opposite of the collector current variation experienced by the bipolar transistor. The mutual conductance of a FET will fall at high frequencies. The cut-off frequency f_c is the frequency at which g_m has fallen to 0.7 times its value at 1000 Hz.

Handling the MOSFET

The gate terminal of a MOSFET is insulated from the channel by a very thin ($\simeq 100$ nm) layer of silicon dioxide, which effectively forms the dielectric of a capacitance. Any electric charge which accumulates on the gate terminal may easily produce a voltage across the dielectric that is of sufficient magnitude to break down the dielectric. Once this happens the gate is no longer insulated from the channel and the MOSFET has been destroyed. The charge necessary to damage a MOSFET need not be large since the capacitance between the gate and channel is very small and voltage = charge/capacitance. This means that a dangerously high voltage can easily be produced by merely touching the gate leads with a finger or a tool. To prevent damage to MOSFETS in store or about to be fitted into a circuit, it is usual for them to be kept with their gate and source leads short-circuited together. The protective short-circuit can be provided by twisting the leads together, by means of a springy wire clip around the leads, or by inserting the leads into a conductive jelly or grease. The short-circuit must be retained in place while the device is fitted into a circuit, particularly during the soldering process.

Some MOSFETS are manufactured with a Zener diode internally connected between the gate and the substrate. Normally, the voltage across the diode is too low for it to conduct and it has little effect on the operation of the device. If a large voltage should be developed at the gate by a static electric charge, the Zener diode will break down and prevent the MOSFET being damaged.

The Field-effect Transistor as a Switch

A FET can be employed as an electronic switch since its drain current can be turned ON or OFF by the application of a suitable gate-source voltage. In the ON condition the gate-source voltage has moved the operating point to the top of the load line (see Fig. 4.17), and maximum drain current flows. The voltage across the FET, known as the *saturation* voltage $V_{DS(SAT)}$, does not fall to zero but is typically in the range 0.2 V to 1.0 V. In the case of the FET characteristics in the figure, the ON resistance is 0.9 V/7.6 mA or 118 Ω but this value is somewhat higher than many other FETS present; typically R_{ON} is some 30–200 Ω. To turn the FET OFF the gate-source voltage is reduced so that the operating point is shifted to the bottom of the load line. The drain current is now reduced to a very small value, typically 1 nA for a JFET and about 50 pA for a MOSFET.

The minimum time taken by a FET to switch from one state to the other is another important feature. Power is mainly dissipated within a FET switch during the time it is passing from one state to the other

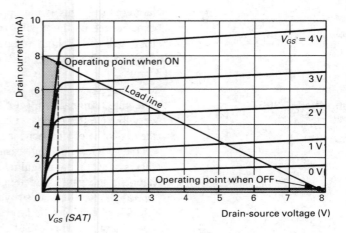

Fig. 4.17   The FET as a switch

since, when the device is ON or OFF, either the voltage across it or the current in it is very small, and power dissipation is the product of voltage and current. The faster the switching speed of a FET the higher is its efficiency.

Enhancement MOSFETs are commonly employed as switches in digital integrated circuits. When large currents are to be switched ON and OFF either a bipolar transistor, a power MOSFET (page 90), or a IGBT (page 93) is normally used.

The FET is not subject to charge storage delay as is the bipolar transistor. Limitation of the switching speed arises because of the presence of inevitable stray capacitances between both the gate-source and the drain-source terminals. The physical structure of a FET produces inter-terminal capacitances; these are the gate-drain capacitance C_{gd}, the gate-source capacitance C_{gs} and the drain-source capacitance C_{ds}. In data sheets these are represented by C_{iss} (input = $C_{gs} + C_{gd}$), C_{oss} (output = C_{ds}) and C_{rss} (reverse transfer = C_{gd}) capacitances. In determining the switching performance of a device C_{iss} is the most important factor. When an input voltage pulse is applied to a FET the input capacitance must be charged, and discharged, before V_{GS} can rise, or fall, to its final value. The lower the source resistance the faster will be the speed of switching.

The turn-on delay time $t_{d(ON)}$ is the time required to charge the input capacitance to the gate threshold value. The risetime t_r is the time required to charge the input capacitance to a specified value above the threshold figure. The turn-off time $t_{d(OFF)}$ is the time needed to discharge the input capacitance from an over-drive voltage to the saturated gate-source voltage. The falltime t_f is the time taken by the gate voltage to fall from the saturated value to the threshold value, and for the output capacitance to charge up to the supply voltage.

Typically, $t_{d(ON)} = 15$ ns, $t_r = 50$ ns, $t_{d(OFF)} = 90$ ns, and $t_f = 50$ ns.

FET Switching Circuits

FETs can be used in analogue circuits to connect a signal voltage to a load on the receipt of a command signal and in digital circuits to switch an output voltage between zero volts and a positive voltage. Fig. 4.18 shows the basic circuit of a series JFET analogue switch. When the control voltage is 0 V the JFET has $V_{GS} = 0$ V and so it is conducting. The JFET is operated below its pinch-off voltage so that its current-voltage characteristic is linear to ensure that changes in signal voltage will not produce a distorted output voltage. When the JFET is conducting its resistance is fairly low, typically 100 Ω, and so the input voltage appears across the load resistance with little reduction in its amplitude. When the control voltage goes negative the JFET turns OFF and then the input signal is unable to pass through the circuit and the output voltage is zero.

An alternative switching circuit is shown by Fig. 4.19. Now the input signal is able to pass through the circuit to the load whenever the control voltage is negative and the JFET is turned OFF. When the control voltage is 0 V the JFET is ON and there is zero output voltage.

An n-channel depletion-mode MOSFET can be turned ON with a positive voltage, or turned OFF with a negative voltage, applied to the gate terminal as shown by Fig. 4.20. Zero gate voltage does not turn the MOSFET OFF. This is a disadvantage and hence depletion-mode devices are less often used as switches than enhancement-mode MOSFETs. An n-channel enhancement-mode MOSFET is turned OFF with zero gate voltage and turned ON by a positive gate voltage. The basic circuit of an n-channel enhancement-mode MOSFET switching circuit is shown by Fig. 4.21.

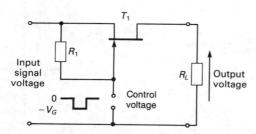

Fig. 4.18 JFET series analogue switch

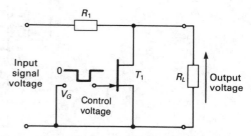

Fig. 4.19 JFET parallel analogue switch

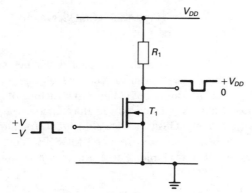

Fig. 4.20 Depletion MOSFET switch

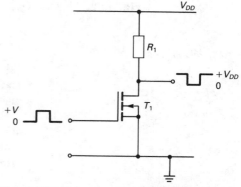

Fig. 4.21 Enhancement MOSFET switch

Enhancement MOSFETS are commonly employed in digital integrated circuits as switches and then the drain resistance is usually provided by another MOSFET. The drain of the MOSFET which is to act as a resistor is connected to its gate as shown by Fig. 4.22. Two enhancement-mode MOSFETs are used in this way in an LSI logic family known as NMOS. This family was often used in the past, but increasingly in modern circuitry both n-channel and p-channel devices are used in what is known as the CMOS logic family. The basic circuit of a CMOS switch is shown by Fig. 4.23. When the control voltage is positive the n-channel T_1 is turned ON and the p-channel T_2 is turned OFF. The output voltage of the circuit is then 0 V. Conversely, when the control voltage is reduced to 0 V T_1 turns OFF and T_2 turns ON; now the output voltage of the circuit is equal to the supply voltage V_{DD}. Often the source terminal of the n-channel MOSFET T_1 is connected to a negative voltage V_{SS} instead of to earth. Then the T_1 ON voltage is $-V_{SS}$ volts.

The use of two MOSFETs, one n-channel and the other p-channel, connected in series between the supply voltage and earth like this has the advantage that when the circuit is in either of its two states one MOSFET is ON and the other is OFF. There is hence very little current taken from the supply and so the power dissipation of the circuit is very small.

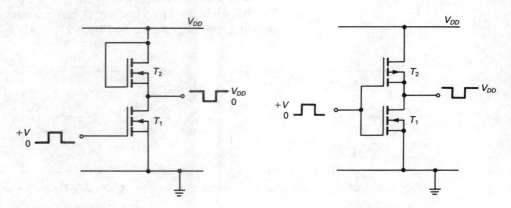

Fig. 4.22 NMOS switch   Fig. 4.23 CMOS switch

Power MOSFETs

A power MOSFET is an n-channel or a p-channel device that is capable of handling a large drain current at a high drain-source voltage and that is particularly suited both to the fast switching of large currents and to audio-/radio-frequency power amplification.

Originally, power MOSFETs were all *vertical structure metal oxide silicon power field-effect transistor* or VMOSFET devices, but now other types are also available. These are known as double-diffused DMOSFETs or TMOSFETs.

A switching device used in power control circuitry should:

(a) have a low ON resistance
(b) be able to switch rapidly from one state to the other
(c) take only a small input current, or better still, be voltage-controlled
(d) when conducting be able to continuously carry the full rated load current
(e) be able to bear the peak value of the supply voltage.

Power MOSFETs satisfy these requirements and they are hence a common choice for use in modern power electronic circuits. Their main disadvantage is that their ON resistance is often bigger than desirable and for this reason they are sometimes replaced by a newer device, the insulated gate bipolar transistor (IGBT).

VMOSFETs

Fig. 4.24 shows the construction of an n-channel VMOSFET. It differs from the other types of FET mainly in the characteristic V-shaped groove and the positioning of the drain terminal at the bottom of the device. The gate terminal is insulated from the rest of the FET by a layer of silicon dioxide or of quartz. The source and drain terminals are placed on opposite sides of the structure so that any flow of drain current is *vertically* from drain to source. This is to be contrasted with the current flow in the JFET and in the MOSFET in which current flows horizontally.

A channel does not exist between the drain and the source until a potential is applied to the gate terminal to make it positive with respect to the source. Then, electrons will be attracted to beneath the two surfaces of the V-shaped groove to induce a channel between the source terminal and the n^- region. Current is then able to flow from the drain to the epitaxial region and thence through the induced channel to the source terminal. The channel is of shorter length than in the

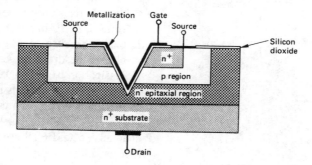

Fig. 4.24 VMOSFET

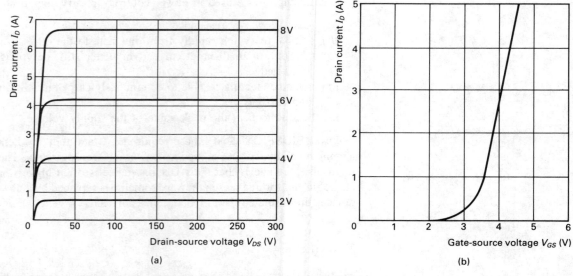

Fig. 4.25 (a) Drain and (b) mutual characteristics of a VMOSFET

more conventional FETS and this reduces both the ON resistance and the self-capacitances of the VMOSFET.

The p-n junctions between the p region and the two n regions are reverse-biased and this results in a depletion region that lies mainly in the n⁻ region. The presence of this high-resistance region gives the VMOSFET a high value of drain-source breakdown voltage. Typical drain and mutual characteristics for a VMOSFET are given in Fig. 4.25. They can be seen to be of similar shape to those for the JFET and MOSFET, but involve much larger currents and voltages.

A number of advantages are claimed for the VMOSFET over the JFET and the MOSFET. These include:

(a) the current density can be much larger
(b) the self-capacitances are smaller
(c) the ON resistance is very much smaller, typical values being some 2 to 10 ohms
(d) partly because of (a) and partly because the construction allows the drain to be connected to a heat sink, the power dissipation of a VMOSFET can be much larger
(e) the switching speed is much faster, typical turn-on and turn-off times being 4 ns each.

P-channel VMOSFETS are also available, although they are not as often used.

DMOSFETs and TMOSFETs

The V-shaped groove in the VMOSFET causes a large electric field to be produced within the FET that has a tendency to break down the insulating layer at the tip of the groove. This is sometimes partially overcome by the use of a flat-bottomed groove but also by employing the alternative construction that is shown by Fig. 4.26.

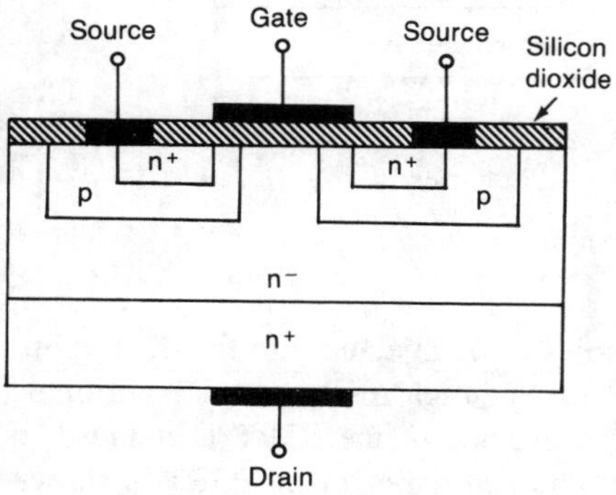

Effectively, two MOSFETs known as *cells*, are connected in parallel and have a common drain terminal. The source cells, which are hexagonal in shape for the DMOSFET and square for the TMOSFET, are isolated from each other and from the common drain by the p regions. The cells share the total current between them. When the gate is made positive relative to the source, enough electrons are attracted into the p-type regions to alter them to be n-type. The source cells are then no longer isolated from the drain and a drain current is able to flow to each source cell. Fig. 4.26 shows only two source cells, but a practical device would employ many thousands of them.

The symbols used for the VMOSFET, the DMOSFET and the TMOSFET are the same as for an enhancement-mode MOSFET.

Fig. 4.26 DMOSFET

The Insulated Gate Bipolar Transistor

The insulated gate bipolar transistor (IGBT) is a MOSFET-controlled switching device which combines the fast operation and high power capability of the bipolar transistor with the voltage control of a MOSFET. The symbol for an IGBT is shown by Fig. 4.27(a). The structure of an IGBT is shown by Fig. 4.27(b). It is similar to the structure of a DMOSFET but it uses a p^+-type substrate instead of an n^+ substrate. The operation of the device is similar to that of an enhancement-mode MOSFET which drives a p-n-p bipolar transistor. The equivalent circuit of the IGBT is given by Fig. 4.28.

The application of a positive gate-emitter voltage forms a channel (as with an enhancement-mode MOSFET), to connect the two n^+ regions either side of the gate. The current that flows through this channel is the base current for the p-n-p transistor. The p^+ regions in each cell together with the aluminium contacts that connect the n^+ and p^+ regions give an effective shunt resistance R_s in each cell. The shunt resistor R_S reduces the current gain of the n-p-n bipolar transistor to prevent the circuit from latching up. These features allow the IGBT to be controlled by the voltage applied to the gate terminal over a wide range of drain currents and voltages.

The device is normally operated with a positive voltage, relative to the emitter, applied to the collector. When the gate-emitter voltage is 0 V the collector current is zero. As the gate-emitter voltage is increased positively electrons pass into the n^- region, which acts as the base of the p-n-p transistor. These electrons forward bias the

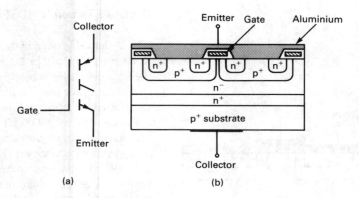

Fig. 4.27 IGBT symbol

(a)   (b)

p^+n^- junction and so cause holes to be injected into the n^- region. As a result the conductivity of the high-resistivity n^- region is increased and hence the ON resistance of the IGBT is reduced to a low value. A typical output characteristic for an IGBT is shown in Fig. 4.29.

Metal Semiconductor Field-effect Transistors

The metal semiconductor field-effect transistor (MESFET) is an n-channel depletion-mode device that employs gallium arsenide as the semiconductor material. Gallium arsenide (GaAs) is employed since electrons are able to travel with velocities of up to five times the velocity of an electron moving in silicon. The much higher electron velocities lead to FETs that have much higher cut-off frequencies and/or are able to switch in a very short time, perhaps just a few picoseconds.

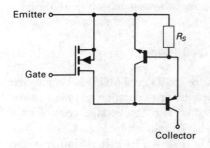

Fig. 4.28 Equivalent circuit of an IGBT

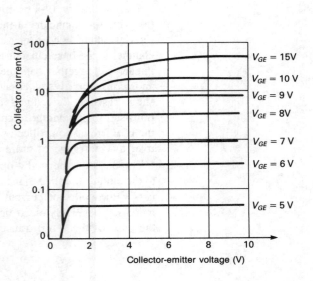

Fig. 4.29 IGBT output characteristic

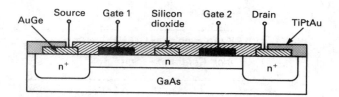

Fig. 4.30 Structure of a MESFET

The structure of a MESFET is shown in Fig. 4.30. The substrate consists of high-resistivity GaAs on to which are formed (i) a thin n-type layer that acts as the channel, and (ii) two n^+ layers that act as the source and drain regions of the FET. On to the surface of the GaAs are deposited two aluminium areas for the two gate contacts, and gold-germanium (AuGe) areas for the drain and source regions, and then lastly silicon dioxide insulation. Each end of the device has a titanium-platinum-gold (TiPtAu) region deposited on to its surface. The aluminium gate contacts are in direct contact with the semi-conductor material so that each acts in the same way as a Schottky diode (not as a p-n junction).

When the gate-source voltage is 0 V, electrons move from the source to the drain through the thin n-type region beneath the gate contact and these provide the drain current. As the gate-source voltage is made more negative the width of the channel is reduced and the drain current falls. Eventually, the gate-source voltage reaches a value at which the channel is closed off and then the drain current is reduced to zero. Fig. 4.31 shows the drain characteristics of a typical MESFET.

The circuit operation of a MESFET is very similar to that of a MOSFET but the device is able to operate at frequencies up to about 20 GHz or, if used as a switch, it is able to switch states in just a few picoseconds.

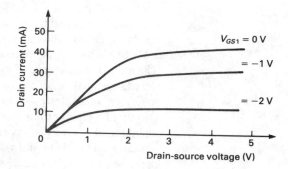

Fig. 4.31 Drain characteristics of a MESFET

Data Sheets

The symbols employed in FET data sheets follow the same system as that used with bipolar transistors. Graphs may be given for such parameters as (a) drain current plotted to a base of drain-source

Table 4.2

BF245 n-channel JFET

Quick Reference Data

$V_{DS(max)}$	30 V	$V_{GSO(max)}$ 30 V		$P_{D(max)}$ 300 mW
I_{DSS}	(V_{DS} = 15 V; V_{GS} = 0 V) > 2 < 6.5 mA			
V_{GS}	(cut-off)	(I_D = 10 nA, V_{DS} = 15 V)	0.5 to 8 V	
g_m	(V_{DS} = 15 V; V_{GS} = 0; f = 1000 Hz;		T_{amb} = 25°C) 3.0 to 6.5 S	

Ratings

$V_{DS(max)}$	30 V	$V_{DGO(max)}$	30 V	$V_{GSO(max)}$	30 V
$I_{D(max)}$	25 mA	$I_{G(max)}$	10 mA	$P_{tot(max)}$	300 mW

Characteristics

V_{GS}	=	20 V;	V_{DS} = 0 V	I_{GSS}	<	5 nA
V_{DS}	=	15 V;	V_{GS} = 0 V	I_{DSS}	=	2 to 6.5 mA
I_G	=	1 μA;	V_{DS} = 0 V	$V_{GSS(BR)}$	>	30 V
I_D	=	200 μA;	V_{DS} = 15 V	V_{GS}	=	0.4 to 2.2 V
g_m	=	3.0	to 8.0 mS			

voltage, (*b*) drain current against gate-source voltage and (*c*) g_m against frequency. Data for the BF 245 is given in Table 4.2.

As for the bipolar transistor, component distributors often provide FET data in a more concise form which makes it easier to select a suitable device for a particular application (Table 4.3).

The Relative Merits of Bipolar Transistors and FETs

The input impedance of a bipolar transistor depends on the d.c. collector current it conducts under quiescent conditions, and for the majority of applications it is somewhere in the region of 1000−3000 ohms. If the transistor is biased so that its quiescent collector current is only a few microamperes, an input impedance of a megohm or more can be achieved. The input impedance of a JFET is very high, with a MOSFET having an impedance which is several orders higher still. The mutual conductance of a bipolar transistor is considerably higher than the mutual conductance of a FET; this means that the bipolar transistor is capable of providing the larger voltage gain. The collector current of a bipolar transistor increases with increase in temperature and thermal runaway is a possibility unless suitable d.c. stabilization circuitry and/or heat sinks are used. The drain current of a FET decreases with increase in temperature and there is no risk of thermal instability.

When a FET is used as a switch its ON resistance is larger than the ON resistance obtainable from a transistor but the switching operation can be carried out in either direction, i.e. the drain and source terminals are interchangeable. On the other hand, the switching

Table 4.3

A.	n-channel JFETs							
Type no.	$P_{D(max)}$ (mW)	$V_{DS(max)}$ (V)	$V_{DG(max)}$ (V)	$V_{GS(max)}$ (V)	$I_{GSS(max)}$ (nA)	g_m (mS)	Max input capacitance (pF)	$I_{DSS(max)}$ (mA)
BF244	360	30	30	30	8	4.5	5	25
MPF102	200	25	25	25	2	1.6	6	20
2N3823	300	30	30	30	0.5	3.5	7	20

B.	n-channel MOSFETs							
BFR84	300	20	30	20	10	15	6	—
BFS28	200	20	20	50	1	13	—	—
BSV81	200	30	30	10	0.001	—	5	—

C.	Power FETs							
Type no.	Type	$P_{D(max)}$ (W)	$V_{DS(max)}$ (V)	$V_{DG(max)}$ (V)	$V_{GS(max)}$ (V)	g_m (ms)	$I_{D(max)}$ (A)	Max input capacitance (pF)
VK1010	VMOS	1	60	60	15	200	0.5	50
IRF150	DMOS	150	100	100	20	10	28	3000
ZVN3315	DMOS	0.63	150	100	20	—	1	28

D.	IGBTs			
Type no.	$P_{D(max)}$ (W)	V_{GE} (threshold) (V)	$I_{C(max)}$ (A)	$V_{CE(max)}$ (V)
IRGB20S	60	3(min)– 5(max)	19	600
IRGB40S	160	3(min)– 5(max)	50	600

speed of the FET is lower than that of the bipolar transistor. This is because the ON resistance of a FET is larger than that of a bipolar transistor and so a FET is unable to charge or discharge the stray capacitances and the input capacitance of the next stage as quickly.

The power MOSFET can switch at a frequency which is at least ten times higher than is achievable by the best bipolar transistors. Further, its ON resistance is low, certainly comparable with bipolar transistors.

Exercises

4.1 Explain how a depletion layer is formed in the channel of a JFET by the applied drain-source voltage and how the channel resistance varies with this voltage for small voltages. What effect stops this happening at higher voltages?

4.2 Label each part of the JFET shown in Fig. 4.32 and say whether it is a p-channel or an n-channel device.

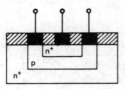

Fig. 4.32

4.3 A JFET has a mutual conductance of 3 mS. A signal of 1.5 V peak is applied between the gate and the source terminals. Calculate the peak drain signal current that flows.

4.4 A JFET has an output resistance of 50 kΩ. When a signal of 1.5 V peak is applied to the transistor the drain current varies by ± 2 mA about its steady (quiescent) value. Calculate the mutual conductance of the transistor.

4.5 When the gate-source voltage of a JFET is held at a constant value it is found that a change in the drain-source voltage of 2 V produces a change in the drain current of 0.5 mA. Calculate the output resistance of the FET.

4.6 Explain how the symbols for the four types of MOSFET indicate (i) whether the device is p-type or n-type channel, (ii) whether drain current will flow in the absence of a gate-source voltage.

4.7 Explain the effects on the performance of a MOSFET of an increase in the temperature of the device.

4.8 Explain the action of a p-channel enhancement-mode MOSFET.

4.9 Fig. 4.33 shows the mutual and drain characteristics of a FET. Calculate, using both characteristics, the mutual conductance when the gate-source voltage is 0.4 V.

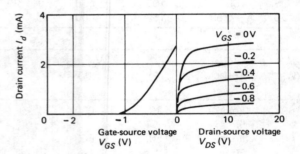

Fig. 4.33

4.10 For the FET quoted in **4.9** calculate its output resistance when $V_{GS} = -0.4$ V.

4.11 For the characteristics shown in Fig. 4.34 calculate the gate-source voltage needed to give a d.c. drain current of 8 mA. If the drain-source voltage is 8 V, estimate the d.c. resistance of the FET for this value of V_{GS}.

4.12 Explain why thermal runaway is not a problem with a FET.

4.13 The drain-source voltage of a JFET is increased from 6 V to 7 V. The resulting increase in the drain current is 0.1 mA. Calculate the output resistance of the FET. Assume that there is zero change in the value of the gate-source voltage.

4.14 A depletion-mode MOSFET has the data given in Table 4.4. Plot the mutual characteristic and hence determine the mutual conductance of the device when the gate-source voltage is 1.2 V.

4.15 In a FET circuit, $V_{gs} = 1.5$ V, $E_s = 1$ V, $R_s = 2500$ ohms and $R_L = 3000$ ohms. If the mutual conductance g_m of the FET is 2 mS calculate the a.c. voltage developed across R_L.

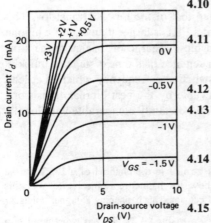

Fig. 4.34

Table 4.4

Drain current (mA)	7.0	5.5	4.0	2.5	1.7	0.5	0
Gate-source voltage (V)	0	0.5	1.0	1.5	2.0	3.0	4.0

4.16 Explain how the channel depletion region of a p-channel JFET increases in width as the drain-source voltage is increased. Make clear whether this voltage increases from zero volts to a positive or a negative value. What polarity must the gate-source voltage be to increase further the width of the depletion region?

4.17 The maximum gate current $I_{G(max)}$ for a particular type of JFET is 200 nA. Calculate its input resistance when $V_{GS} = 2$ V.

4.18 Fig. 4.35 shows the maximum, typical and minimum mutual characteristics of an n-channel JFET. Determine the maximum, typical and minimum value of the mutual conductance g_m of the FET.

4.19 The drain characteristics of a FET are given by the data of Table 4.5. Plot the characteristics and state which type of FET it is. Calculate the values of the mutual conductance g_m and the drain-source resistance r_{ds} when $V_{DS} = 6$ V.

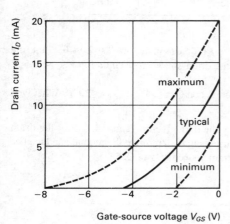

Fig. 4.35

Table 4.5

	Drain current I_D (mA)					
Drain-source voltage V_{DS}(V)	Gate-source voltage V_{GS}(V)					
	−3	−2	−1	0	+0.5	+1
2	0.8	2.5	5.0	7.7	8.6	11.8
4	0.8	3.0	5.8	8.6	10.5	13.0
8	0.8	3.1	5.9	9.0	11.4	13.5
20	0.8	3.2	6.0	9.2	11.8	14.0

4.20 Use the data given in Table 4.5 to plot the mutual characteristic of the FET for $V_{DS} = 8$ V.

4.21 Fig. 4.36 shows the drain characteristics of an n-channel enhancement-mode MOSFET which is to be used as a switch. Mark the characteristics with the ON and OFF regions of the device. Estimate the saturation voltage $V_{DS(SAT)}$.

4.22 When a FET is ON its drain current is 8 mA. If $V_{DS(SAT)} = 0.8$ V calculate the ON resistance of the FET.

4.23 What is meant by the active region of a FET characteristic? Why does a FET dissipate more power when it is in this region than when it is in either its ON or its OFF region?

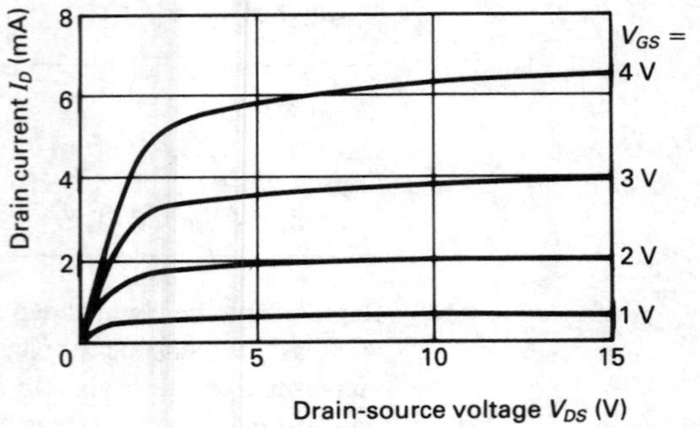

Fig. 4.36

Drain-source voltage V_{DS} (V)

Table 4.6

Drain-source voltage V_{DS}(V)	Drain current I_D (mA)				
	Gate-source voltage V_{GS}				
	1 V	0.5 V	0 V	−0.5 V	−1 V
10	4.00	3.19	2.38	1.57	0.76
20	4.02	3.21	2.40	1.59	0.78
30	4.04	3.23	2.43	1.61	0.80

4.24 The drain characteristics of a FET are given in Table 4.6. Plot the characteristics and determine the drain-source resistance for $V_{GS} = 0.5$ V from the characteristic.

Use the curves to find the mutual conductance for $V_{DS} = 20$ V. What type of FET is this?

4.25 An n-channel junction FET has the data given in Table 4.7. Plot the drain and mutual characteristics and use them to determine the mutual conductance of the device. Find also the drain-source resistance.

Table 4.7

Drain-source voltage V_{DS}(V)	Drain current I_D (mA)			
	Gate-source voltage V_{GS}			
	0 V	−0.5 V	−1.0 V	−1.5 V
10	2.25	1.35	0.7	0.3
20	2.29	1.38	0.73	0.33
30	2.32	1.41	0.75	0.35

5 Integrated Circuits

The methods used to fabricate silicon planar bipolar and field-effect transistors can be extended to allow a complete circuit to be manufactured in a single silicon chip. All the components, active and passive, which are required by the circuit are formed at the same time in a small piece of silicon, known as a *chip*. The circuit is known as a monolithic integrated circuit or IC because only one silicon chip is used. The use of monolithic integrated circuits has a number of advantages over the use of discrete circuits: greatly reduced size and weight; lower costs; complex circuit functions are economically possible, e.g. pocket calculators; and greater reliability. The size and weight reductions occur because a quite complex circuit can be enclosed within a volume of comparable dimensions to those of a single transistor. The cost of an integrated circuit depends on its complexity and the quantity manufactured, but in many cases the cost is no greater than that of one transistor.

Other types of integrated circuit, known as thin-film and thick-film circuits, are also available; with both thin-film and thick-film circuits, resistors and capacitors are fabricated by forming a suitable film on to the surface of a glass or a ceramic substrate. The components are interconnected in the required manner by means of a deposited metallic pattern. Thin-film components are produced by vacuum deposition of a suitable material on to the surface of the substrate. Thick-film components are produced by painting the substrate with special kinds of ink. Active components and inductors cannot be produced in this way and any such components that are necessary must be provided in discrete form, and be joined into the metallic pattern at the appropriate points. Thick- and thin-film circuits are not used to anywhere near the same extent as monolithic circuits and they will not be discussed in this book.

Integrated Circuit Components

The fabrication of an integrated circuit component is achieved by a sequential series of oxidizing, etching and diffusion. The components that can be formed by this process are transistors (both bipolar and field-effect), diodes, resistors and capacitors; inductors cannot be produced.

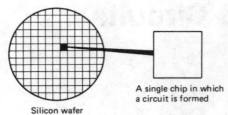

Silicon wafer

A single chip in which
a circuit is formed

Fig. 5.1 Showing how a silicon wafer is divided into a number of chips

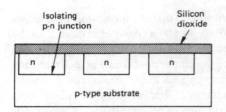

Isolating
p-n junction

Silicon
dioxide

n   n   n

p-type substrate

Fig. 5.2 Method of isolating the components in an integrated circuit

A thin wafer, about 5–10 mils* thick, is sliced from a rod of p-type silicon and will have a surface area of about 4 in². Since an integrated circuit may only occupy an area of about 30 mils², several thousands of identical circuits can be simultaneously fabricated in the one wafer. The principle is illustrated by Fig. 5.1, although to simplify the drawing fewer chips have been shown.

Each individual silicon chip acts as a substrate into which the various components making up the circuit can be formed. The components are simultaneously formed by the diffusion of impurity elements into selected parts of the chip.

Since the p-type silicon substrate is an electrical conductor it is necessary to arrange that each of the components is insulated from the substrate. If this is not done the various components will all be coupled together by the substrate resistance. There are a number of different ways in which the required isolation can be obtained, but the most common method utilizes the high-resistance property of a reverse-biased p-n junction (see Fig. 5.2). Several n-type regions, equal in number to the number of components in the circuit, are diffused into the p-type substrate. Each of the n-type regions will be isolated from the substrate if the junction is maintained in the reverse-bias condition by connecting the substrate to a potential which is more negative than any other part of the circuit.

The various components making up the circuit are fabricated by means of a number of n-type and p-type regions which are diffused into the isolated regions, see Fig. 3.33. Once formed, the components are interconnected as required by the circuit by means of an aluminium pattern deposited on to the surface of the chip.

Integrated Circuit Bipolar Transistor

The construction of an n-p-n bipolar transistor is shown in Fig. 5.3 (n⁺ denotes a region of greater conductivity). The construction is similar to that of the silicon planar transistor, but differs from it in that the collector contact is brought out at the top of the transistor instead of at the bottom. The change in the position of the collector contact is necessary because the collector current cannot be allowed to flow in the substrate. The collector current must therefore flow in the narrow collector region and so the device has a greater collector resistance than the discrete transistor. This, undesirable, series resistance can be minimized by the use of a buried layer. The buried layer consists of an n⁺ low-resistance region diffused into the chip in the position shown in Fig. 5.4. The buried layer is effectively in parallel with the collector region and it reduces the collector series resistance. The series resistance cannot be reduced by using a lower resistivity material for the collector region since this would reduce

* One mil is one thousandth of an inch.

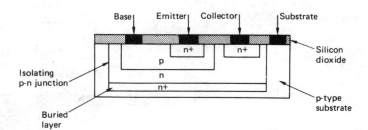

Fig. 5.3 An integrated circuit n-p-n bipolar transistor

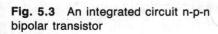

Fig. 5.4 An integrated circuit n-p-n bipolar transistor with a buried layer

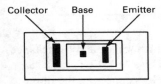

Fig. 5.5 Top view of an IC transistor

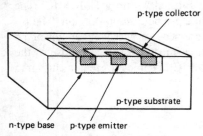

Fig. 5.6 Lateral p-n-p bipolar transistor

the breakdown voltage of the collector-base junction. Typically, an h_{fe} value of about 100 is achieved. At high frequencies, the capacitance of the isolating p-n junction may possess a sufficiently low reactance to couple the collector to the substrate and adversely affect the frequency response. The top view of a bipolar transistor is shown by Fig. 5.5.

The fabrication of a p-n-p transistor is not as simple or as cheap because additional p-type and n-type regions are required. Alternatively, a different and less efficient layout known as a *lateral transistor* can be employed which is more expensive and which provides a lower current gain of about 5 (see Fig. 5.6). Because of the difficulties associated with the use of the p-n-p transistor, its use in an integrated circuit is avoided as much as possible.

Integrated Circuit MOSFET

Most of the MOSFETs employed in digital ICs are n-channel enhancement mode devices. Depletion-mode MOSFETs are not often used and when they are they are generally connected to act as a resistor. P-channel MOSFETs are mainly used in with n-channel devices to form complementary pairs in the CMOS logic family.

The fabrication of an n-channel integrated circuit MOSFET is shown by Fig. 5.7. The MOSFET has an advantage over the bipolar transistor in that it is self-isolating; the drain and source regions are each isolated from the substrate by their individual p-n junctions, while the gate terminal is isolated by a layer of silicon dioxide. This feature allows a MOSFET to be formed in a smaller area of the chip than can be achieved with a bipolar transistor. Because of this MOSFETs are employed in all ICs that include a large number, perhaps many

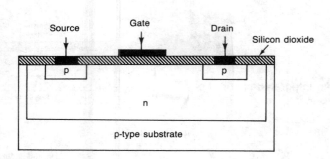

Fig. 5.7 Integrated circuit n-channel enhancement-mode MOSFET

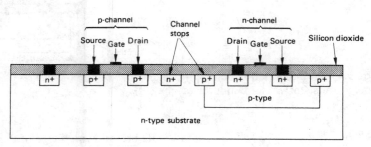

Fig. 5.8 CMOS construction

thousands, of transistors. Also, required resistance and capacitance values, if not too large, are often provided by a MOSFET, capacitance by a reverse-biased junction, resistance by connecting the drain to the gate.

Fig. 5.8 shows how a pair of MOSFETs, one an n-channel and the other a p-channel device, are formed within a CMOS IC. The n-channel FETs are formed within a p-type isolation region. The p^+ and n^+ regions, labelled as *stops*, are necessary to avoid an unwanted channel being induced between adjacent transistors. Because of the need to insert these stops a CMOS IC occupies a greater chip area than an NMOS IC.

Integrated Circuit JFET

The JFET is mainly used in some linear integrated circuits for its high input impedance. Fig. 5.9 shows the construction of an integrated n-channel JFET.

Integrated Circuit Diode

Fig. 5.10 shows the construction of an integrated circuit diode. The diode is formed at the same time as one of the junctions of a transistor and consists of a p-type region (the anode) and an n-type region (the cathode). An n^+ region is diffused into the chip to reduce the resistance of the cathode contact.

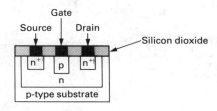

Fig. 5.9 Integrated n-channel JFET

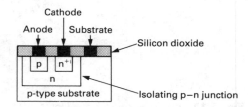

Fig. 5.10 An integrated diode

Integrated Circuit Resistor

Integrated resistors are made using a thin layer of p-type silicon that is diffused at the same time as the base of the transistor. The resistance of a silicon layer depends on the length l, area a and resistivity ρ of the layer according to

$$R = \frac{\rho l}{a} \tag{5.1}$$

The area a of the layer is the product of the width W and the depth d of the layer. Thus

$$R = \rho l / W d \ \Omega$$

It is usual to express the resistance in terms of the resistance of a square of the silicon layer (Fig. 5.11) in which the width W of the layer is equal to the length l. Then, equation (5.1) can be written as

$$R = \frac{\rho l}{ld} = \frac{\rho}{d} \ \Omega / \square \tag{5.2}$$

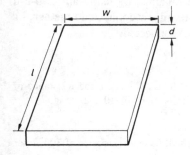

Fig. 5.11

The resistance is now the resistance between the opposite sides of a square and it is measured in a unit known as the ohm per square. The resistance depends only on the resistivity of the silicon layer and not on the dimensions of the square. The resistivity of the layer is determined by the number of charge carriers (holes) that are diffused into the layer and the depth to which they penetrate. However, since both of these variables are fixed by the requirements of the simultaneously diffused transistors, a required resistance value must be obtained by a suitable choice of the length and width of the resistive path. The resistance value given by a square can be increased by increasing the length of the path. Difficulties are experienced with the fabrication of very high values of resistance because of the relatively large chip area such resistances demand.

The constructional details of an integrated resistor are given in Figs. 5.12(a) and (b). Fig. 5.12(a) shows that the resistive path is formed by a p-type region that joins together the resistor contacts. A p-type path is used since it will be diffused at the same time as transistor

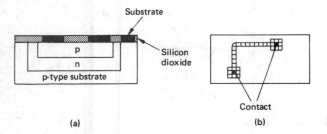

Fig. 5.12 (a) Side and (b) top views of an integrated resistor

base regions and will therefore be only lightly doped. The resistivity will therefore be in the range of 100–300 $\Omega/\square$. When very low values of resistance are required an n-type resistor is employed; the n-type path is diffused at the same time as the transistor emitter regions and will therefore be of much lower resistivity. Fig. 5.12(b) shows the top view of an integrated resistor and indicates how a required resistance value may be obtained by connecting in series a number of 'squares'. The resistor can follow any path that will best utilize the surface area of the chip. The practical range of resistance values is from about 15 Ω to about 30 kΩ.

The tolerance of an IC resistor is rather poor, being of the order of 20%. Resistors formed in the same chip can be closely matched, certainly to within about 1%. Since IC resistors take up a relatively large chip area their use within an IC is avoided as much as possible. A required resistance can often be obtained by using an IC n-p-n transistor with its base and emitter terminals connected together, or a FET with its gate and source interconnected.

Example 5.1

The resistivity of a p-type region is 100 $\Omega/\square$. Calculate the resistance of a strip which is 1 mil wide and (i) 20 mils long (ii) 30 mils long.

Solution
Since the resistive strip is 1 mil wide it will have a resistance of 100 Ω per 1 mil length.

(i) The resistance of a 20 mil length is $20 \times 100 = 2000\ \Omega$ (*Ans.*)
(ii) The resistance of a 30 mil length is $30 \times 100 = 3000\ \Omega$ (*Ans.*)

Example 5.2

An IC resistor is made up from a length of 4 mil wide resistive strip of resistivity 50 $\Omega/\square$. Calculate its resistance if the strip is (a) 18 mils (b) 19 mils long.

Solution
(a) The resistor is equivalent to a 16 mil length of 4 mil wide strip plus two 2 × 2 mil squares, as shown by Fig. 5.13(a). The resistance of each is the same regardless of its dimensions. Hence, the equivalent circuit is as

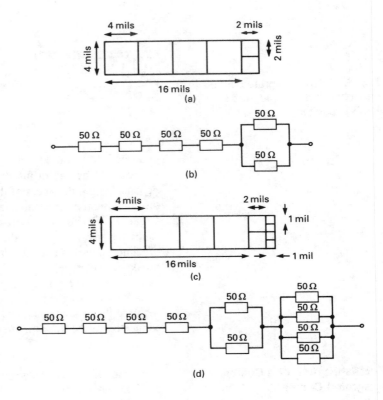

Fig. 5.13

shown by Fig. 5.13(b). From this circuit the resistance R of the IC resistor is $4 \times 50 + 50/2 = 225 \ \Omega$ (Ans.)

(b) The 19 mil long resistor is shown in Figs. 5.13(c) and (d). It effectively consists of four 4×4 mils squares plus two 2×2 mil squares plus four 1×1 mil squares. The resistance R is

$$R = 4 \times 50 + 50/2 + 50/4 = 237.5 \ \Omega \quad (Ans.)$$

Integrated Circuit Capacitor

Integrated capacitors can be fabricated in two ways: either the capacitance of a reverse-biased p-n junction can be utilized, or the capacitance can be provided by a layer of silicon dioxide separating two conducting areas. The construction of a junction-type capacitor is shown in Fig. 5.14(a). The p-n junction is formed at the same time as either the emitter-base or the collector-base junction of a transistor. Provided the p-n junction is held in the reverse-biased condition, a capacitance of about 0.2 pF/mil can be obtained. Since the area of the chip available for a capacitor is limited, values of up to about 100 pF are available. Fig. 5.14(b) shows a MOS capacitor; one electrode of the capacitor is provided by an aluminium layer that is

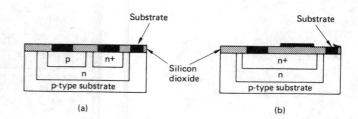

Fig. 5.14 (*a*) Integrated circuit capacitor and (*b*) integrated circuit MOS capacitor

deposited on to the top of the silicon layer and the other electrode is produced by the diffused n^+ region. The capacitance provided depends on the thickness of the silicon dioxide layer and the area of the aluminium plate; up to a few hundred picofarads can be achieved.

The MOS capacitor is more expensive to provide but has the following advantages over the junction capacitor: it can have voltages of either polarity applied to it, it has lower losses and a larger breakdown voltage, and its capacitance value does not depend on the magnitude of the voltage applied across the capacitor.

Large values of capacitance cannot be provided within an IC; if required the capacitance must be supplied by an external component connected to the appropriate terminals of the IC.

The Fabrication of a Complete Integrated Circuit

The main differences between integrated and discrete circuits which perform the same function are that the integrated circuit uses transistors and diodes as liberally as possible. This is because resistors and capacitors occupy more space in the chip than do transistors and they are therefore more expensive.

In the fabrication of a complete integrated circuit all the components, active and passive, required to make up the circuit are formed at the same time. The components are then interconnected as required by means of an aluminium pattern which is deposited on the top of the silicon slice. As an example, suppose the simple circuit shown in Fig. 5.15(*a*) is to be integrated. Fig. 5.15(*b*) shows the three components of the circuit diffused into a p-type substrate. The components are each isolated from the substrate by a reverse-biased p-n junction and are connected together in the required manner by an aluminium pattern which is deposited on to the surface of the chip.

The problem of interconnecting the various components in an IC increases with increase in the number of components. Fig. 5.16 shows the top view of an IC that contains ten bipolar transistors, six diodes and six resistors. There are also six terminals to be brought out of the package that can be used for the connection of external components and power supplies. A metallization pattern must be deposited on the top of the chip to form all of the required interconnections between the components. Clearly, it can be difficult to work out the best pattern and a computer with the appropriate software is often used to aid in this task.

A large number of circuits are simultaneously produced in a single

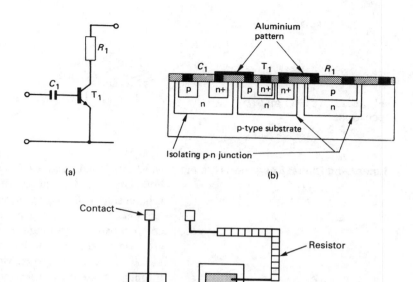

(a)

(b)

(c)

Fig. 5.15 Showing (a) a simple transistor circuit, (b) the same circuit in integrated form and (c) the layout of the chip (top view)

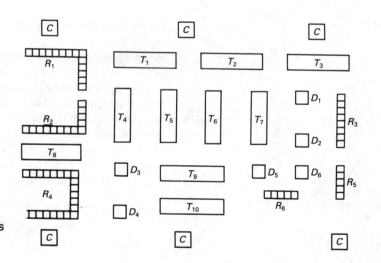

Fig. 5.16 A complex circuit requires metallized interconnections

silicon wafer, and after formation they are separated into individual chips and then sealed within a suitable package. The majority of integrated circuits are available in one or more of two kinds of package; these packages are the TO circular packages and the dual-in-line (DIL). The two packages are illustrated by Figs. 5.17(a) and (b), the latter being much the more popular.

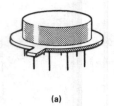 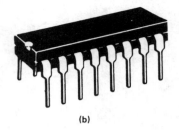

Fig. 5.17 (a) TO and (b) DIL packages

(a)   (b)

Linear and Digital Integrated Circuits

Most integrated circuits (ICs) are classified as being either *linear* or *digital* devices. A linear circuit is one whose output signal has a linear relationship with the input signal. Linear ICs are also known as *analogue* ICs because the signals that they handle may vary continuously over a range of values. Linear ICs include operational amplifiers (op-amps), audio-frequency power amplifiers and various radio and television circuits. Most linear ICs employ bipolar transistors throughout, although a few types of op-amp have either a JFET or a MOSFET input stage.

Digital ICs operate with binary digital signals that are always in either of two possible states; namely logic 0 and logic 1 (see page 164). Digital ICs include gates, counters, memories and micro-processors. Some ICs include both analogue and digital circuitry; three examples of these ICs are timers, analogue-to-digital converters (ADC) and digital-to-analogue converters (DAC).

Linear Integrated Circuits

When a linear integrated circuit is used in electronic circuitry a number of external components are also necessary and must be connected to the appropriate terminals of the IC. An operational amplifier, or *op-amp*, is a high-gain voltage amplifier that has two input terminals and one output terminal. There are a large number of different op-amps on the market made by a variety of manufacturers, but one of the most commonly used is the 741. The pin connections of this op-amp are shown by Fig. 5.18(a) and the symbol for an op-amp is shown by Fig. 5.18(b). One of the inputs is labelled as the inverting terminal

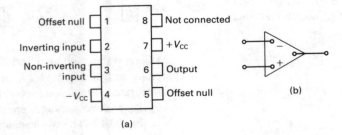

Fig. 5.18 (a) pin connections of the 741 op-amp and (b) symbol for an op-amp

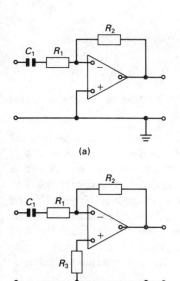

(a)

(b)

Fig. 5.19 (*a*) inverting amplifier and (*b*) R_3 gives d.c. balance to the circuit

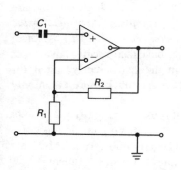

Fig. 5.20 Non-inverting amplifier

and the other input is labelled as the non-inverting terminal. A signal applied to the inverting terminal appears at the output terminal inverted, i.e. a sinusoidal input signal will experience 180° phase shift, but a signal applied to the non-inverting terminal will not be inverted. Most op-amps require both positive and negative power supply voltages, although a few types have been designed to work from a single polarity supply.

When an op-amp is used as a voltage amplifier it must have two resistors connected between the appropriate terminals to specify the voltage gain. Fig. 5.19(*a*) shows the circuit of an inverting amplifier using an op-amp. One resistor R_1 is connected between the input terminal of the circuit and the inverting terminal of the op-amp, and the other resistor R_2 is connected between the inverting input terminal and the output terminal of the op-amp. The voltage gain A_v of the inverting amplifier is

$$A_v = -R_2/R_1 \tag{5.3}$$

The input resistance of the amplifier circuit is equal to R_1 ohms.

Example 5.3

An inverting op-amp circuit is to have a voltage gain of 30 and an input resistance of 2000 Ω. Determine values for the resistors R_1 and R_2.

Solution
The input resistance is to be 2000 Ω and hence $R_1 = 2000$ Ω   (*Ans.*)
From equation (5.3), $R_2 = 30 \times 2$ kΩ $= 60$ kΩ   (*Ans.*)

The 60 kΩ could be obtained using two 120 kΩ resistors in parallel, or the nearest preferred value of 62 kΩ could be used.

A third resistor, R_3 in Fig. 5.19(*b*), is often used to d.c. balance the circuit (essential if there is no input coupling capacitor). The value of R_3 should be equal to the total resistance of R_1 and R_2 in parallel, i.e. $R_3 = R_1R_2/(R_1 + R_2)$.

If a non-inverting voltage gain is wanted, the circuit shown in Fig. 5.20 is used. The voltage gain A_v of the circuit is

$$A_v = (R_1 + R_2)/R_1 \tag{5.4}$$

and the input resistance is very high. If d.c. balance of the circuit is wanted, a resistor R_3 is connected in series with the non-inverting terminal.

Example 5.4

A non-inverting op-amp circuit is to have a voltage gain of 30 and an input resistance of 2000 Ω. Determine the component values.

Solution

From equation (5.4), $30 = (R_1 + R_2)/R_1$
$$29R_1 = R_2$$

Choosing a suitable value for R_2, say 2 kΩ, $R_1 = 58$ kΩ (*Ans.*)
The nearest preferred value is 56 kΩ.

To obtain an input resistance of 2 kΩ, a 2 kΩ resistor can be connected across the input terminals of the circuit.

An audio-frequency power amplifier IC will also need some external resistors and it will require at least one external decoupling capacitor. (A decoupling capacitor is one that is connected in parallel with a resistor to prevent an a.c. voltage being dropped across the resistor. The capacitor has a low reactance at the frequencies of operation and effectively short-circuits the resistor at those frequencies.) Variable resistors, such as volume and tone controls, are also provided externally. Any inductors that may be needed in an analogue circuit cannot be provided within an IC and will also be externally connected.

Digital Integrated Circuits

Digital integrated circuits are all classified as being one of the following: (i) a small-scale integrated circuit (SSI), (ii) a medium-scale integrated circuit (MSI), (iii) a large-scale integrated circuit (LSI) and (iv) a very-large-scale integrated circuit (VLSI). These categories are based on the number of transistors within each IC. An SSI IC has up to 10 transistors, an MSI has between 10 and 100 transistors, LSI ICs have between 100 and 5000 transistors and, lastly, a VLSI circuit contains more than 5000 transistors.

Digital ICs are manufactured using different technologies and are members of various *logic families*. The three basic logic families are (*a*) *transistor-transistor logic (TTL)*, (*b*) *complementary metal oxide semiconductor (CMOS)* and (*c*) *emitter-coupled logic (ECL)*. All three logic families contain sub-families which have their relative advantages and disadvantages making one or another particularly suited for different applications. TTL and CMOS are much more commonly used than ECL.

ECL is the fastest logic family and it is used for high-speed circuits, such as mainframe computers, only. For all other digital circuits either TTL or CMOS is used. Standard TTL and 4000 series CMOS are the earliest versions of these logic families but neither of them is used for new designs now, although devices in these families are still available from manufacturers. TTL has developed first into *low-power Schottky TTL* and then into *advanced Schottky TTL* and *advanced low-power Schottky TTL*. The low-power versions offer a smaller power dissipation and the Schottky versions (which use Schottky diodes internally) offer a much increased speed of operation . The 4000 series CMOS devices are relatively slow to operate but have the advantage of a very small power dissipation. The pin connections of some common TTL and CMOS gates are given on page 172. The speed disadvantage of 4000 series CMOS has been overcome, while retaining its low power dissipation feature, with the modern replacements known as high-speed CMOS and advanced CMOS. For LSI and VLSI circuits either CMOS or NMOS is used, since devices using these technologies occupy less space in a silicon wafer and this factor allows many more devices to be packed into a given chip area. This is important with many of the complex circuits used today, such as microprocessors and large-capacity memories, which often contain many thousands of transistors.

Exercises

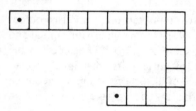

Fig. 5.21

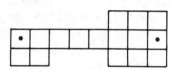

Fig. 5.22

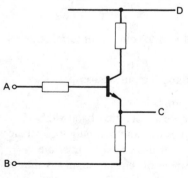

Fig. 5.23

5.1 Show that the resistance/□ of a silicon chip layer is given by $R = \rho/d \ \Omega/\square$, where ρ is the resistivity of the silicon and d is the thickness of the layer.
   A silicon layer has a resistance of 180 $\Omega/\square$. Calculate the resistance of a strip that is 1.5 mil wide and 30 mil in length.

5.2 The resistivity of a silicon layer of sides a and b, where $a = b$, is 120 $\Omega/\square$. What will the resistance be if (a) a is doubled and b is unchanged, (b) both a and b are doubled, (c) a is unchanged and b is doubled?

5.3 Calculate the resistance of the IC resistor shown in Fig. 5.21 if (a) $R = 120 \ \Omega/\square$ and (b) $R = 200 \ \Omega/\square$.

5.4 Calculate the resistance of the IC resistor shown in Fig. 5.22 if $R = 250 \ \Omega/\square$. Suggest another method of obtaining the same resistance value.

5.5 An IC resistor is to be designed using material having a resistivity of 150 $\Omega/\square$. Draw a possible layout if the required resistance is (a) 225 Ω and (b) 375 Ω.

5.6 The capacitance of a junction capacitor in an IC is 0.25 pF/mil². Calculate the chip area needed to provide a capacitor of 50 pF.

5.7 Why are high values of resistance and capacitance not fabricated in an IC chip? How are any such components that are needed provided?

5.8 Explain why an integrated p-n-p transistor cannot be made using the same technique as employed to make an n-p-n transistor.

5.9 Draw the integrated circuit version of the circuit given in Fig. 5.23.

5.10 Very often a required resistance value is obtained in a silicon chip by means of a suitably biased bipolar or field effect transistor. Show how each transistor can be connected to act as a resistance.

5.11 Give four advantages of integrated circuits over their discrete component versions.

5.12 Describe, using appropriate sketches, how either (i) an enhancement-type MOSFET or (ii) a bipolar transistor can be fabricated in integrated circuit form.

5.13 Explain, with sketches, how: (i) a capacitor, (ii) a resistor, (iii) a diode can be formed on a silicon chip. Describe any method which can be used to isolate electrically each component formed within a silicon chip.

5.14 Explain the meaning of the following when applied to integrated circuits: (i) planar process, (ii) diffusion, (iii) epitaxial layer, (iv) metallization.

6 Small-signal Audio-frequency Amplifiers

An amplifier is a circuit that has two input terminals and two output terminals as shown by Fig. 6.1(*a*). Very often one input terminal and one output terminal are common, see Fig. 6.1(*b*). The action of the circuit is to increase the amplitude of, or amplify, the signal applied to the input terminals. In the absence of distortion the input and output signal will have the same waveshape. The voltage gain A_v of an amplifier is defined as

$$A_v = V_{out}/V_{in} \tag{6.1}$$

An audio-frequency amplifier is designed to work over a frequency range of about 25 Hz to 20 kHz and its gain should be more or less constant over this bandwidth.

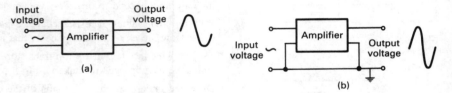

(a)

(b)

Fig. 6.1 Principle of an amplifier: (*a*) separate input and output terminals and (*b*) one input terminal is common with one output terminal

The requirements for a small-signal amplifier are to:

(*a*) have a specified voltage gain
(*b*) provide this voltage gain over a specified bandwidth
(*c*) have specified constant values of input and output resistance
(*d*) introduce as little distortion of the signal as possible
(*e*) work correctly over a range of power supply voltages, particularly for a battery-operated circuit.

A typical specification for an amplifier would be:

voltage gain 200 over a bandwidth of 30 Hz to 15 kHz
input resistance 20 kΩ and output resistance 1 kΩ
distortion less than 0.1%.

Principles of Operation

Bipolar transistors and FETs may be used as amplifiers because their output current can be controlled by an a.c. signal applied to their input terminals. If a voltage or power output is required the output current must be passed through a resistive load. A FET has such a high input resistance that its input current is negligible; it can therefore give only a voltage gain. The input resistance of a bipolar transistor depends on the magnitude of the collector current it is passing and it can be made fairly large if the circuit is biased so that the collector current is only a few microamps. By suitable choice of the collector current, and hence of the input resistance, a bipolar transistor may be considered to be either a current-operated or a voltage-operated device. If the source resistance is much larger than the input resistance of the transistor, the transistor is current operated; if the input resistance is much the larger, it is voltage operated.

Voltage operation of a transistor has the advantage that less noise is introduced. Current operation gives a greater stage gain and, because a larger collector current is employed, a larger maximum permissible output signal. Voltage operation is generally restricted to the input stage of a multi-stage amplifier.

The mutual characteristics of a FET or a bipolar transistor always exhibit some non-linearity, but if a suitable operating point is chosen and the amplitude of the input signal is limited, the operation of the circuit may be taken as linear without the introduction of undue error.

The function of a *small-signal amplifier* is to supply a current or voltage to a load, the power output being unimportant. In a *large-signal amplifier*, on the other hand, the power output is the important factor and to obtain an adequate power output the output current and voltage swings must cover most of the characteristics.

In modern electronics most amplifiers are either op-amps or IC a.f. power amplifiers.

Choice of Configuration

The various ways in which a bipolar transistor or a FET may be connected to provide a gain are shown in Fig. 6.2.

A transistor connected as a common-base amplifier (Fig. 6.2(c)) has a current gain h_{fb} less than unity (typically about 0.992), a low input resistance of the order of 50 Ω, and an output resistance of about 1 MΩ. Because the current gain is less than unity, common-base stages cannot be connected in cascade unless transformer coupling is used. Audio-frequency transformers, however, have the disadvantages of being relatively costly, bulky and heavy and having a limited frequency response.

The current gain h_{fe} of a transistor connected in the common-emitter configuration (Fig. 6.2(a)) is much greater than the current gain of the same transistor connected with common base, i.e. $h_{fe} = h_{fb}/(1 - h_{fb})$. Coupling of the cascaded stages of a multi-stage

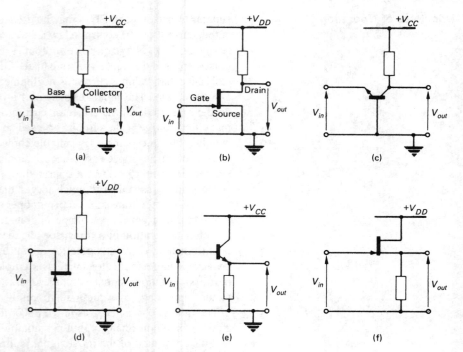

Fig. 6.2 Possible connections for bipolar transistors and FETs

amplifier can be achieved without the use of a transformer. Usually, common-emitter stages are biased so that the transistor is current operated. Then the input resistance is in the region of 1000−2000 Ω while the output resistance is some 10−30 kΩ.

The common-collector circuit, or emitter follower is shown in Fig. 6.2(e). This connection has a high input resistance, a low output resistance and a voltage gain of less than unity. The main use of an emitter follower is as a buffer circuit that can be used to connect a high-resistance source to a low-resistance load.

In the normal mode of operation of a FET amplifier (Fig. 6.2(b)) the source is common to the input and output circuits, the input signal is applied to the gate, and the output is taken from between the drain and earth. This connection provides a voltage gain and it has a high input resistance.

Fig. 6.2(f) shows the FET equivalent of the emitter follower and it is known as the source follower circuit. Lastly, Fig. 6.2(d) shows the common-gate connection; this is not used at audio frequencies.

Most a.f. small-signal amplifiers employ the bipolar transistor because of its much larger mutual conductance, and hence larger voltage gain.

Choice of Operating Point

The mutual characteristic of a FET with a resistive drain load shows how its drain current varies with change in its gate-source voltage for a particular value of drain supply voltage. Similarly, the mutual characteristic of a bipolar transistor shows how the collector current of the transistor varies with change in the base-emitter voltage for a particular value of collector supply voltage.

A mutual characteristic can be used to determine, graphically, the waveform of the output current for a particular input signal waveform. Ideally, the two waveforms should be identical, but this requires the mutual characteristic to be absolutely linear. In practice, some non-linearity always exists and, for minimum signal distortion, care must be taken to restrict operation to the most linear part of the characteristic. For this a suitable *operating* or *quiescent point* must be selected and the amplitude of the input signal must be limited, so that the operation is kept to the more linear part of the characteristic. The chosen operating point is determined by the application of a steady bias voltage or current. For the maximum signal handling performance the operating point is usually placed in the centre of the linear portion of the mutual characteristic. Then an alternating signal centred on this operating point produces equal swings of output current above and below the quiescent value, as shown in Figs. 6.3(a) and (b) for a bipolar transistor.

The output current may conveniently be regarded as a direct current having an alternating current superimposed on it. The direct current is equal to the current that flows when the input signal is zero, known as the quiescent current, and the alternating current has a peak-to-peak value of $I_{max} - I_{min}$.

The output current clearly flows at all times during each cycle of the input signal waveform. The active device is said to be operated under *Class A* conditions. The peak value of the signal waveform should, at all times, be less than the bias voltage, or current, otherwise the output waveform will be distorted.

Note that Fig. 6.3(b) shows the input current as having a sinusoidal waveform; however, the relationship between the input voltage and the input current of a bipolar transistor is not a linear one unless the steady bias current is much larger than the signal current. This is

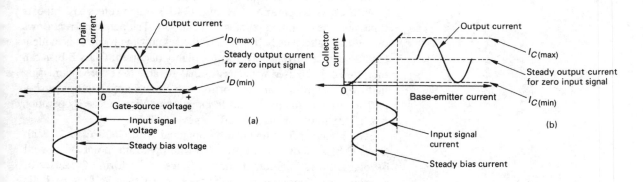

Fig. 6.3 Variations of output current with input signal for (a) a FET and (b) a transistor

because the input resistance of a transistor is not a constant quality but varies with change in input voltage.

The selection of the operation point for a Class A amplifier depends on a number of factors:

(a) *Maximum output voltage*. If the maximum possible output voltage is wanted from a transistor amplifier stage the collector-emitter voltage V_{CE} should be chosen to have a value which is approximately one-half of the collector supply voltage V_{CC}. This choice will position the operating point roughly half-way along the d.c. load line drawn on the output characteristics.

(b) *Maximum voltage gain*. To obtain the maximum voltage gain the h_{fe} of the transistor should be at, or near, its maximum value. This means that the transistor should be biased to have a particular value of collector current. Also, the collector resistor should be of as high a value as possible.

(c) *Minimum noise*. When noise is an important consideration the bias point must be such that the collector current is very small.

(d) *Battery-operated equipment*. When the circuit is a part of a battery-operated equipment a most important consideration will be keeping the current taken from the battery as low as possible. This means that the quiescent collector current should be kept as small as possible. Often the chosen operating point is a compromise between these conflicting factors.

Small-signal a.f. amplifiers always operate under Class A conditions.

Class B and Class C Operation

Class A operation of an amplifier offers low signal distortion. The maximum theoretical efficiency with which the d.c. power taken from the power supply can be converted into a.c. signal power output is, however, only 50%, and practical efficiencies are lower than this. To obtain a greater efficiency than 50%, an amplifier may be operated under either Class B or Class C conditions.

With *Class B* operation (see Fig. 6.4(a) which refers to a bipolar transistor) the operating point is set at cut-off. The output current flows only during alternate half-cycles of the signal waveform. It is evident that the output current waveform is highly distorted; Class B bias can therefore only be used with circuits that are able to restore the missing half-cycles of the signal waveform. Such circuits are known as *push-pull* amplifiers and *tuned radio-frequency* amplifiers. Class B operation has a maximum theoretical efficiency of 78.5%.

Even greater efficiency can be obtained with *Class C* bias. With Class C bias, shown in Fig. 6.4(b), the operating point is set well beyond cut-off. The output current flows in the form of a series of narrow pulses having a duration which is less than half the periodic time of the input signal waveform and it is therefore highly distorted.

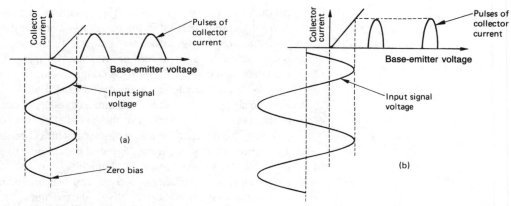

Fig. 6.4   (*a*) Class B, (*b*) Class C operation of a transistor

Class C bias is used with radio-frequency power amplifiers and with some oscillator circuits. These employ a tuned collector load that is able to reject all the unwanted frequencies that are contained in the amplified signal and so restore the original signal waveshape. Class C bias is only used for amplifiers that handle sinusoidal signals.

Another type of operation is known as Class D. In a Class D amplifier the output transistors are operated as switches that are either ON or OFF. The required output voltage is obtained by varying the mark-space ratio of a rectangular wave. The power dissipation of a Class D amplifier is low because the output transistors are either ON or OFF and so the efficiency is very high.

Bias and Stabilization

To establish the chosen operating point it is necessary to apply a bias voltage or current to a FET or transistor.

Transistor Bias

If the current flowing into the base of a common-base-connected transistor is reduced to zero, only the collector leakage current still flows. When there is an input current the d.c. collector current is the sum of the amplified d.c. emitter current and the leakage current, i.e.

$$I_C = h_{FB}I_E + I_{CBO} \tag{6.2}$$

I_{CBO} may be only 10 nA for a silicon planar transistor and perhaps a few microamperes for a germanium transistor.

When a transistor is connected in the common-emitter configuration the base current becomes the input current and the d.c. current gain is increased to h_{FE}. Then

$$I_C = h_{FE}I_B + I_{CEO} \tag{6.3}$$

An increase in the temperature of the collector-base junction will produce an increase in I_{CBO}. The resulting increase in collector current gives an increase in the power dissipated at the collector-base junction, and this, in turn, increases still further the temperature of the junction to give a further increase in I_{CBO}. The process is cumulative, and particularly in the common-emitter connection (since $I_{CEO} \gg I_{CBO}$), leads to signal distortion caused by the operating point moving along the load line. In extreme cases the eventual destruction of the transistor may occur; this unwanted effect is known as *thermal runaway*. To prevent thermal runaway it is often necessary to employ a bias circuit that gives some degree of d.c. stabilization. The current gain h_{FE} and base-emitter voltage V_{BE} are also functions of temperature and variations in them can lead to further changes in collector current. In addition, individual transistors of a given nominal h_{FE} may have values of h_{FE} lying between quoted maximum and minimum values. For example, one transistor's data sheet quotes h_{FE} in the range 125–500.

An amplifier stage will be designed to have a particular value of d.c. collector current using the nominal value of h_{FE}. The bias circuit should operate to ensure that approximately the same collector current will flow if a transistor using either the maximum or the minimum h_{FE} values should be used or if the temperature should change.

Fixed Bias

The simplest method of establishing the *operating point* of a common-emitter transistor is shown in Fig. 6.5. Applying Kirchhoff's second law to the circuit,

$$V_{CC} = I_B R_1 + V_{BE}$$

where V_{BE} is the base-emitter voltage of T_1. Therefore

$$R_1 = (V_{CC} - V_{BE})/I_B \qquad (6.4)$$

This circuit does not provide any d.c. stabilization against changes in collector current due to change in I_{CBO} or in h_{FE} and so its usefulness is limited to 'one-off' circuits.

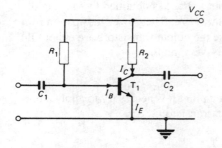

Typical values
$R_1 = 560\,\text{k}\Omega$, $R_2 = 2.7\,\text{k}\Omega$
$C_1 = C_2 = 0.47\,\mu\text{F}$

Fig. 6.5 Fixed bias

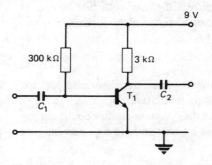

Fig. 6.6

Example 6.1

The circuit shown in Fig. 6.6 is designed for operation with transistors having a nominal h_{FE} of 100. Calculate the d.c. collector current. If the range of possible h_{FE} values is from 50 to 160, calculate the d.c. collector current flowing if a transistor having the maximum h_{FE} is used. Assume $I_{CBO} = 10\,\text{nA}$ and $V_{BE} = 0.62\,\text{V}$.

Solution
From equation (6.4)

$$I_B = (9 - 0.62)/(300 \times 10^3) = 27.9\,\mu\text{A}$$

From equation (6.3)

$$I_C = h_{FE}I_B + I_{CEO} = h_{FE}I_B + (1+h_{FE})I_{CBO}$$
$$= (100 \times 27.9) + (101 \times 10 \times 10^{-3})\ \mu\text{A}$$

Therefore

$$I_C = 2.791\ \text{mA} \qquad (Ans.)$$

Using a transistor of $h_{FE} = 160$,

$$I_C = (160 \times 27.9) + (161 \times 10 \times 10^{-3})\mu\text{A} = 4.46\ \text{mA} \qquad (Ans.)$$

The effect of the increased collector current would be to move the operating point along the d.c. load line (from $I_C = 2.79$ mA to $I_C = 4.46$ mA), and this would lead to signal distortion unless the input signal level were reduced. Because both h_{FE} and V_{BE} vary with change in temperature, the operating point will also shift with any change in temperature.

Collector-base Bias

A better bias arrangement, shown in Fig. 6.7, is to connect a bias resistor R_1 between the collector and base terminals of the transistor. The circuit provides some degree of d.c. stabilization against changes in the designed-for value of the collector current. An increase in the collector current produces an increased voltage drop across the collector load resistor R_2. This causes the collector-emitter voltage to fall, and since this voltage is effectively applied across the base resistor R_1, the base bias current falls also. The fall in bias current leads to a fall in the collector current which to some extent compensates for the original increase.

When a signal voltage is applied to the input terminals of the circuit an amplified version of the signal appears at the collector of T_1. This voltage is fed back, via R_1, to the base circuit. The fed-back voltage will reduce the voltage gain of the circuit — an effect known as *negative feedback*. Should negative feedback not be required the bias circuit will be decoupled by capacitor C_1 as shown by Fig. 6.8.

Potential-divider Bias

For an improvement in the d.c. stabilization the bias arrangement of Fig. 6.9 may be employed. The base of the transistor is held at a positive potential V_B by the potential divider $(R_1 + R_2)$ connected across the collector supply $(V_B = V_{CC}R_2/(R_1 + R_2))$, and the emitter is held at a positive potential V_E by the voltage developed across the emitter resistor R_4. The emitter voltage is then 0.6 V less positive than the base, i.e. $V_E = V_B - 0.6$, and so the emitter current is given by $V_E/R_4 = (V_E - 0.6)/R_4$. If the current gain of the transistor is fairly large the collector current is very nearly equal to the emitter current, $I_C = I_E$, and the voltage drop across the collector load resistor is $I_E R_3$. This means that the collector current is not dependent on the d.c. current gain h_{FE} of the transistor.

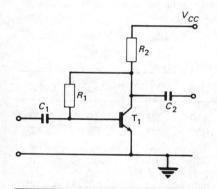

Typical values
$R_1 = 330\,\text{k}\Omega,\ R_2 = 5.6\,\text{k}\Omega$
$C_1 = 1\,\mu\text{F},\ C_2 = 10\,\mu\text{F}$

Fig. 6.7 Collector-base bias

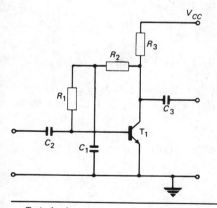

Typical values
$R_1 = 100\,\text{k}\Omega,\ R_2 = 220\,\text{k}\Omega$
$R_3 = 3.3\ \text{k}\Omega,\ C_1 = C_2 = C_3 = 4.7\,\mu\text{F}$

Fig. 6.8 Collector-base bias decoupled to prevent negative feedback

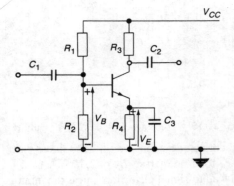

Typical values
$R_1 = 82$ kΩ
$R_2 = 15$ kΩ
$R_3 = 3.3$ kΩ
$R_4 = 1$ kΩ
$C_1 = C_2 = 10 \mu F$
$C_3 = 25 \mu F$

Fig. 6.9 Potential-divider bias

D.C. stabilization of the collector current is achieved in the following manner: an increase in the d.c. collector current, caused by an increase in the temperature of the collector-base junction, is accompanied by an almost equal increase in the emitter current. This results in an increase in the voltage V_E developed across the emitter resistor, and this in turn reduces the forward bias of the emitter-base junction. The base current is reduced causing a decrease in the collector current that compensates for the original increase.

The variations in h_{FE} affect only the base bias current. To prevent these variations affecting the operating point of the circuit the current flowing in R_2 must be several times larger than the base bias current. Typically, the current in R_2 is ten times larger than the base current.

Table 6.1 gives other possible sets of component values as alternatives to those given in Fig. 6.9.

Table 6.1

R_1(kΩ)	R_2(kΩ)	R_3(kΩ)	R_4(Ω)	C_1(μF)	C_2(μF)	C_3(μF)
68	15	2.2	560	10	10	100
120	33	3.3	1000	10	22	47
47	8.2	3.3	820	22	22	100
82	20	3.8	1200	10	50	10
100	18	4.7	1200	4.7	10	22

Example 6.2

In Fig. 6.9 the collector supply voltage V_{CC} is 12 V and the collector current I_C is 1.24 mA. Calculate (a) V_{CE} and (b) V_{BE}.

Solution

(a) $V_{CE} = V_{CC} - I_C R_4 - I_C R_3$
$= 12 - [(1.24 \times 10^{-3}) (3300 + 1000)]$
$= 12 - 5.33 = 6.67$ V   (Ans.)

(b) $V_E = I_C R_4 = 1.24 \times 10^{-3} \times 1000 = 1.24$ V
$V_B = V_{CC} \times R_2/(R_1 + R_2) = 12 \times 15/(15 + 82) = 1.86$ V

Therefore,

$V_{BE} = 1.86 - 1.24 = 0.62$ V   (Ans.)

Example 6.3

A circuit of the type shown in Fig. 6.9 has the component values given in the bottom row of Table 6.1. If the collector current I_C is 1.4 mA and V_{CE} = 5 V calculate (a) the collector supply voltage V_{CC}, and (b) the power dissipated in R_3.

Solution

(a) $V_{CC} = V_{CE} + I_C (R_3 + R_4)$
$= 5 + (1.4 \times 10^{-3}) (1200 + 3800) = 12$ V (*Ans.*)

(b) $P = I_C^2 R_3 = 1.4^2 \times 10^{-6} \times 3800 = 7.45$ mW (*Ans.*)

When a signal voltage is applied to the input terminals of the circuit the base-emitter voltage is caused to vary and this variation produces changes in the collector current. The a.c. component of the collector current passes through the collector resistance and develops an a.c. signal voltage across it. This voltage is then applied, via coupling capacitor C_2, to the output terminal of the amplifier.

The a.c. component of the emitter current will also develop a signal voltage across the emitter resistor that will be in anti-phase with the input signal voltage. This voltage will subtract from the input signal voltage and effectively reduce the gain of the circuit. This is an effect known as *negative feedback* that is sometimes employed in amplifier circuits.

Example 6.4

A voltage source of e.m.f. 50 mV and internal resistance 1 kΩ is connected to the input terminals of the circuit given in Fig. 6.9. The transistor has h_{fe} = 100 and h_{ie} = 2 kΩ. The output coupling capacitor is connected to a load of 1 kΩ resistance. Calculate the signal current supplied to the load.

Solution

The bias resistors and the transistor's h_{ie} are effectively in parallel with one another. Hence the input conductance of the circuit is

$1/R_{in} = (1/15 + 1/82 + 1/2) \times 10^{-3} = 0.579$ mS, and $R_{in} = 1.7$ kΩ
$V_{in} = (50 \times 1.7)/(1 + 1.7) = 31.48$ mV
$I_b = V_{in}/h_{ie} = (31.48 \times 10^{-3})/(2 \times 10^3) = 15.7$ μA
$I_c = h_{fe}I_b = 100 \times 15.7 \times 10^{-6} = 1.57$ mA

The effective collector load resistance $= (3.3 \times 1)/(3.3 + 1) = 0.767$ kΩ

$V_{out} = 1.57 \times 0.767 = 1.2$ V, and therefore
$I_{load} = 1.2/1000 = 1.2$ mA (*Ans.*)

Bandwidth

The gain-frequency characteristic of an audio-frequency amplifier is shown in Fig. 6.10. It can be seen that the voltage gain is constant over a wide frequency band and falls at both low and high frequencies. The *bandwidth* of an amplifier is the band of frequencies over which the gain is not less than $1/\sqrt{2}$ times the maximum gain.

The input coupling capacitor C_1 is required to prevent the base bias current being affected by the resistance of the source. The function of the coupling capacitor C_2 is to prevent d.c. current taken from the collector supply flowing into the load. The values of C_1 and C_2 are

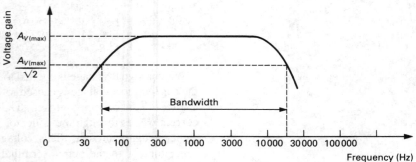

Fig. 6.10 Gain-frequency characteristic of an audio-frequency amplifier

chosen to ensure that they have negligible reactance at most of the frequencies at which the circuit is to operate. At the lower frequencies the reactances of C_1 and C_2 will increase, and some of the signal voltage will be dropped across them. This means that the voltage gain of the circuit falls with decrease in frequency at those low frequencies at which the reactances of C_1 and C_2 are not negligibly small.

The increased reactance of decoupling capacitor C_3 at low frequencies will mean the emitter resistor R_4 is decoupled inadequately. This will allow some *negative feedback* to develop and still further reduce the voltage gain.

The gain at high frequencies will decrease with increase in frequency because of unavoidable circuit capacitances which effectively shunt the signal path. Further loss of gain will occur if the current gain of the transistor falls with increase in frequency, but this effect can be easily avoided by choosing a transistor with a sufficiently high f_t.

Example 6.5

The voltage gain of an amplifier is constant at 220 over the frequency band 40 Hz to 8 kHz. At 20 Hz and at 10 kHz the gain has fallen to 110. Plot the gain-frequency characteristic of the amplifier and from it determine the bandwidth of the amplifier.

Solution

The characteristic is shown plotted in Fig. 6.11. The bandwidth of the amplifier is the band of frequencies between the two points on the characteristic at which the gain is equal to $220/\sqrt{2} = 156$. From the characteristic the bandwidth is $9200 - 28 = 9172$ Hz (*Ans.*)

Gain-frequency distortion of a signal will occur when the gain of an amplifier is not constant with frequency. The output signal waveform will then differ from the input signal waveform because the different frequency components of the signal are amplified to different extents.

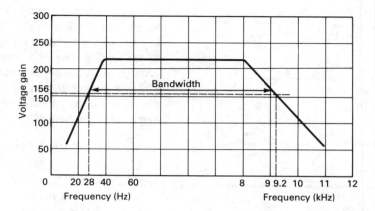

Fig. 6.11

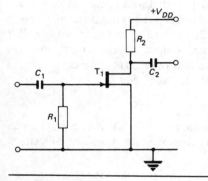

Typical values
$R_1 = 2.2$ MΩ, $R_2 = 5.6$ kΩ
$C_1 = 0.1 \mu$F, $C_2 = 10 \mu$F

Fig. 6.12 JFET simple bias

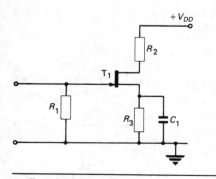

Typical values
$R_1 = 2$ MΩ, $R_2 = 7.8$ kΩ
$R_3 = 4.7$ kΩ, $C_1 = 22 \mu$F

Fig. 6.13 JFET and depletion-type MOSFET source bias

FET Bias

The drain characteristics shown in Fig. 4.13 show that a JFET is conducting when the gate-source voltage V_{GS} is zero. The simplest method of biasing an n-channel JFET is therefore that given in Fig. 6.12. The disadvantages are (a) the maximum input signal amplitude must be very small if excessive distortion is to be avoided, and (b) no stabilization against changes in the d.c. drain current is provided.

Normally an n-channel JFET is operated with its gate biased negatively with respect to its source. This can be achieved by the circuit shown in Fig. 6.13. Resistor R_1 connects the gate to the earth line and the voltage drop across R_3 provides the required bias voltage. The gate current is minute and hence, for values of R_1 of a megohm or so, the d.c. voltage developed across R_1 is negligibly small. Resistor R_3 is often decoupled by means of capacitor C_1 to prevent negative feedback of the signal. This arrangement provides adequate d.c. stability for most small-signal stages provided that the temperature variation is not greater than about 20°C from room temperature. JFETs of the same type are subject to wide spreads in some of their parameters and it may therefore often be necessary to use the more effective bias circuit of Fig. 6.14, which operates in similar fashion to the circuit shown in Fig. 6.9.

Improved stabilization of the drain current can be achieved if another resistor R_2 is connected between the drain supply voltage and the gate terminal (Fig. 6.14). R_1 and R_2 form a potential divider across the drain supply voltage to keep the gate potential constant. The gate-source voltage V_{GS} is the difference between the potentials of the gate and the source. If the drain current should increase for some reason, the voltage across R_3 will increase and this will make the gate potential more negative relative to the source potential. V_{GS} will then be more negative and the drain current will fall, tending to compensate for the original increase in its value.

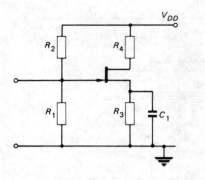

Typical values
$R_1 = 56\,k\Omega$, $R_2 = 120\,k\Omega$,
$R_3 = 1\,k\Omega$, $R_4 = 4.7\,k\Omega$,
$C_1 = 22\,\mu F$

Fig. 6.14 Method of biasing a JFET

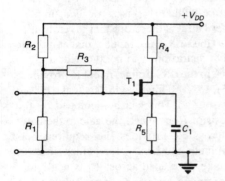

Typical values
$R_1 = 12\,k\Omega$, $R_3 = 1.2\,M\Omega$
$R_2 = 10\,k\Omega$, $R_4 = 4.7\,k\Omega$
$R_5 = 3.9\,k\Omega$, $C_1 = 22\,\mu F$

Fig. 6.15 JFET and depletion-type
MOSFET potential-divider bias

Typical values
(a) $R_1 = 300\,k\Omega$, $R_2 = 3.9\,k\Omega$
(b) $R_1 = 200\,k\Omega$, $R_2 = 100\,k\Omega$
    $R_3 = 3.9\,k\Omega$

Fig. 6.16 Enhancement-type
MOSFET bias

If the junction of the bias resistors R_1 and R_2 is directly connected to the gate terminal of the FET then, at all signal frequencies, the bias resistors will effectively appear in parallel with the input terminals of the device. The high input resistance of the FET will then be reduced to a considerably smaller value. To minimize this shunting effect, a resistor R_3 can be used to connect the bias resistors to the gate terminal as shown in Fig. 6.15. The input resistance of the amplifier is then equal to

$$R_3 + R_1R_2/(R_1 + R_2) \tag{6.5}$$

R_3 is chosen to be 1 MΩ or more and hence the input resistance is approximately R_3.

An n-channel depletion-mode MOSFET must be biased so that its gate is held at a negative potential relative to its source and hence either of the bias circuits shown in Figs. 6.14 and 6.15 can be employed. An n-channel enhancement-mode MOSFET must be operated with its gate at a positive potential with respect to its source and so a different bias circuit is necessary. The circuit shown in Fig. 6.16(a) can be used if the operating point $V_{GS} = V_{DS}$ is suitable. If, for reason of obtaining maximum output voltage or minimizing distortion, some other operating point is required then the circuit given in Fig. 6.16(b) must be used. With this circuit

$$V_{GS} = V_{DS}R_2/(R_1 + R_2) \tag{6.6}$$

Both circuits provide d.c. stabilization of the operating point in a similar manner to that previously described for transistor collector-base bias.

Example 6.6

In the circuit of Fig. 6.16(b) the operating point is to be at $I_D = 2$ mA, $V_{GS} = 2.5$ V, and $V_{DS} = 4$ V. If the drain supply voltage is 9 V determine suitable values for the three resistors.

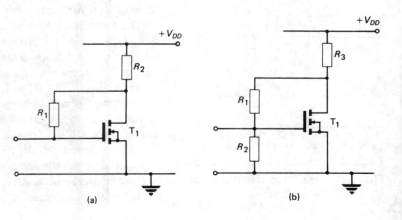

(a)

(b)

Solution

$$I_D R_3 = 9 - 4 = 5 \text{ V}$$
$$R_3 = 5/(2 \times 10^{-3}) = 2.5 \text{ k}\Omega. \text{ Nearest preferred value}$$
$$= 2.4 \text{ k}\Omega \quad (Ans.)$$
$$2.5 = 4R_2/(R_1 + R_2), \ 1 + R_1/R_2 = 4/2.5 = 1.6$$
$$R_1 = 0.6R_2$$

The chosen values should be fairly large to keep a high input resistance.

If $R_2 = 200 \text{ k}\Omega$ then $R_1 = 120 \text{ k}\Omega$ (*Ans.*)

Determination of Gain using a Load Line

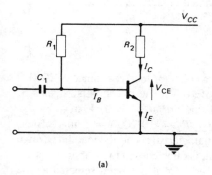

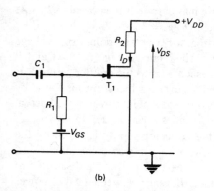

Fig. 6.17 Currents and voltages in basic amplifiers

The voltage gain of a FET amplifier or the current gain of a bipolar transistor amplifier can be determined with the aid of a *load line* drawn on the output current-output voltage characteristics of the device. The currents and voltages existing in the collector or drain circuit of a simple resistance-loaded amplifier are shown in Figs. 6.17(*a*) and (*b*) respectively. In each of these circuits the d.c. collector or drain current flows in the load resistor R_2 and develops a voltage across it. The direct voltage which is applied across the transistor or FET is equal to the supply voltage minus the d.c. load voltage.

Thus, referring to Fig. 6.17,

$$V_{CE} = V_{CC} - I_C R_2 \tag{6.7}$$

$$V_{DS} = V_{DD} - I_D R_2 \tag{6.8}$$

Equations (6.6) and (6.7) are of the form $y = mx + c$ and are therefore equations of a straight line. In order to draw a straight line it is only necessary to plot two points; these points can best be determined in the following manner. Point A: Let $I_C = I_D = 0$ in equations (6.7) and (6.8) respectively, then

$$V_{CE} = V_{CC} \quad \text{and} \quad V_{DS} = V_{DD}$$

Point B: Let $V_{CE} = V_{DS} = 0$ in equations (6.7) and (6.8) respectively, then

$$0 = V_{CC} - I_C R_2 \quad \text{or} \quad I_C = V_{CC}/R_2$$

$$0 = V_{DD} - I_D R_2 \quad \text{or} \quad I_D = V_{DD}/R_2$$

If the points A and B are marked on the characteristics and then joined together by a straight line, the line drawn is the *load line* for the particular values of load resistance and supply voltage. The load line can be used to determine the values of current and voltage in the output circuit, since the ordinate of the point of intersection of the load line and a given input current or voltage curve gives the output current or voltage for that input signal.

Consider for example, the output characteristics of an n-p-n transistor given in Figs. 6.18(*a*) and (*b*), and suppose the transistor

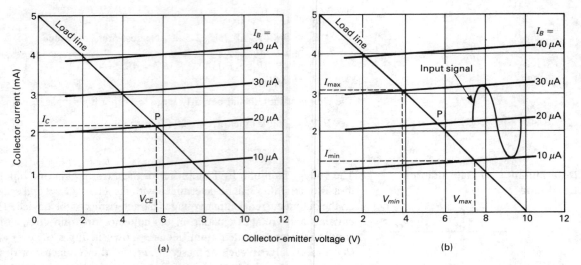

Fig. 6.18

is to be used in an amplifier with a collector load resistance of 2000 Ω and a collector supply voltage of 10 V. The two points, A and B, that locate the ends of the load line are at $I_C = 0$, $V_{CE} = V_{CC} = 10$ V, and at $V_{CE} = 0$, $I_C = V_{CC}/R_2 = 10/2000 = 5$ mA. These points have been located on the characteristics and the load line drawn between them. The operating point is often chosen to lie approximately in the middle of the load line, and here it has been selected as the point marked P. The required base bias current is then equal to 20 μA. The d.c. collector-emitter voltage ($V_{CE} = V_{CC} - I_C R_2$) is found by projecting vertically downwards from the operating point to the voltage axis. This step is shown by a dashed line in Fig. 6.18(a) and it determines the standing (or quiescent) collector-emitter voltage as 5.6 V. Similarly, the d.c. collector current which flows is found by projecting horizontally from the operating point towards the current axis. Thus, the d.c. collector current is equal to 2.2 mA.

The d.c. power taken from the collector supply is given by the product of the collector supply voltage and the d.c. collector current, i.e. $10 \times 2.2 \times 10^{-3} = 22$ mW. The d.c. power P_C dissipated at the collector of the transistor is the difference between the d.c. power supplied to the circuit and the d.c. power dissipated in the load resistance. Hence

$$P_C = 22 \times 10^{-3} - [(2.2 \times 10^{-3})^2 \times 2000] = 12.32 \text{ mW}$$

P_C is also equal to the product of the d.c. collector current and the quiescent collector-emitter voltage; thus

$$P_C = 2.2 \times 10^{-3} \times 5.6 = 12.32 \text{ mW}$$

The load line can also be used to find the variations in both the collector current and the collector-emitter voltage which are produced by the application of a signal to the base of the transistor. Suppose as shown by Fig. 6.18(b) that a sinusoidal signal of peak value 10 μA is applied to the transistor. This signal is superimposed on the base bias current of 20 μA and so the base current is varied from a minimum value of 10 μA to a maximum value of 30 μA. The corresponding values of the collector current and the collector-emitter voltage are determined by projecting to the current and voltage axes from the points of intersection of the load line and the curves for $I_B = 10$ μA and $I_B = 30$ μA. This has been shown by the dashed lines drawn on Fig. 6.18(b). The collector current is varied from a minimum value $I_{C(min)} = 1.3$ mA to a maximum value $I_{C(max)} = 3.25$ mA. The collector-emitter voltage is varied from a minimum value $V_{CE(min)} = 3.6$ V to a maximum value $V_{CE(max)} = 7.3$ V. The a.c. component of the collector current has a peak-to-peak value of $(3.25 - 1.3)$ mA or 1.95 mA, while the peak-to-peak value of the a.c. component of the collector-emitter voltage is $(7.3 - 3.6)$ or 3.7 V.

The current gain A_i of the amplifier is

$$A_i = \frac{\text{Peak-to-peak change in collector current}}{\text{Peak-to-peak change in base current}}$$

Hence

$$A_i = (1.95 \times 10^{-3})/(20 \times 10^{-6}) = 97.5$$

Example 6.7

The transistor used in a single-stage audio-frequency amplifier with a resistance load of 2000 Ω has the data given in Table 6.2.

Plot the output characteristics of the transistor and draw the load line assuming a collector supply voltage V_{CC} of 8 V.

(i) Select a suitable operating point.
(ii) Determine the current gain A_i when an input signal producing a base current swing of 5 μA about the chosen bias current is applied to the circuit.
(iii) Assuming the input resistance of the transistor is 1900 Ω determine the voltage gain A_v.
(iv) Calculate the power gain A_p.

Table 6.2

V_{CE}(V)	I_C(mA)			
	$I_B = 5$ μA	$I_B = 10$ μA	$I_B = 15$ μA	$I_B = 20$ μA
2	0.85	1.55	2.32	3.08
4	1.00	1.74	2.56	3.35
6	1.13	1.92	2.76	3.60
8	1.30	2.13	3.00	3.85

Solution

The output characteristics are shown plotted in Fig. 6.19. The d.c. load line must be drawn between the points

$$I_C = 0, \quad V_{CE} = V_{CC} = 8 \text{ V}, \quad \text{and}$$
$$V_{CE} = 0, \quad I_C = V_{CC}/R_2 = 8/2000 = 4 \text{ mA}$$

(i) Since the input signal has a peak value of ± 5 μA a suitable base bias current is 10 μA, the operating point is then P.

(ii) When a signal of ± 5 μA peak is applied to the transistor, the base current varies between 5 μA and 15 μA. Projection from the intersection of the load line and the 5 μA and 15 μA base current curves to the current axes gives the resulting values of collector current as 1.15 mA and 2.45 mA.

The peak-peak collector current swing is therefore 2.45 – 1.15 or 1.30 mA and the current gain is

$$A_i = (1.3 \times 10^{-3})/(10 \times 10^{-6}) = 130 \quad (Ans.)$$

(iii) If the input resistance of the transistor is 1900 Ω the a.c. voltage applied to the transistor must be

$$\pm 5 \times 10^{-6} \times 1900 \quad \text{or} \quad \pm 9.5 \times 10^{-3} \text{ V}$$

Projecting from the intersection of the load line and the appropriate base current curves to the voltage axis gives the peak-peak collector voltage as

$$5.7 - 3.08 = 2.62 \text{ V}$$

Voltage gain $A_v = 2.62/(19 \times 10^{-3}) = 138 \quad (Ans.)$

Alternatively, using equation (3.7),

$$A_v = A_i R_L/R_{in} = (130 \times 2000)/1900 = 137 \quad (Ans.)$$

(iv) The power output of the transistor is the product of the r.m.s. values of the a.c. components of the collector current and the collector-emitter voltage. Therefore

$$\text{Output power} = (\text{peak-peak } I_c)/2\sqrt{2} \times (\text{peak-peak } V_{ce})/2\sqrt{2}$$

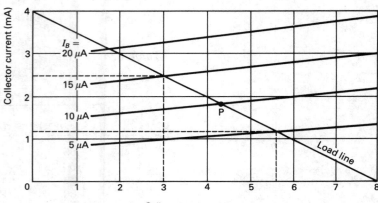

Fig. 6.19

$$= [(I_{c(max)} - I_{c(min)}) \; (V_{ce(max)} - V_{ce(min)})]/8$$
$$= (1.3 \times 10^{-3} \times 2.62)/8 = 4.258 \times 10^{-4} \text{ W} \quad (Ans.)$$

The input power delivered to the transistor is

$$I^2{}_{b(rms)} h_{ie} = [(5 \times 10^{-6})/\sqrt{2}]^2 \times 1900 = 23.75 \times 10^{-9} \text{ W}$$

Power gain $A_p = P_{out}/P_{in} = 4.258 \times 10^{-4}/23.75 \times 10^{-9} = 17\,928$
$(Ans.)$

Alternatively, using equation (3.8),

$$A_p = A_i^2 R_L/R_{in} = A_i A_v = 130 \times 137 = 17\,810 \quad (Ans.)$$

Example 6.8

Determine the mutual conductance of the transistor in Example 6.4. Use this value to calculate the voltage gain of the circuit.

Solution

$$g_m = h_{fe}/h_{ie} = 130/1900 = 68.4 \text{ mS} \quad (Ans.)$$
$$\text{or, } g_m = 38 \times 10^{-3} \times I_C = 38 \times 10^{-3} \times 1.8 = 68.4 \text{ mS} \quad (Ans.)$$
$$\text{Voltage gain} = g_m R_L = 68.4 \times 10^{-3} \times 2 \times 10^3 = 137 \quad (Ans.)$$

Since the mutual conductance g_m is primarily determined by the d.c. collector current its value is more predictable than is the value of h_{fe} and so g_m is increasingly employed in performance calculations.

Example 6.9

The bipolar transistor used in an audio-frequency amplifier has $h_{fe} = 250$ and $R_{in} = 1500 \; \Omega$. (*a*) Calculate the collector load resistance needed to give a voltage gain of 200. (*b*) A voltage source of e.m.f. 20 mV and internal resistance 600 Ω is connected to the input terminals of the amplifier. Calculate the output voltage.

Solution

(*a*) $200 = 250 R_L/1500$, $R_L = (200 \times 1500)/250 = 1200 \; \Omega \quad (Ans.)$
(*b*) $V_{in} = (20 \times 1500)(600 + 1500) = 14.29 \text{ mV}$
$V_{out} = 200 \times 14.29 \times 10^{-3} = 2.86 \text{ V} \quad (Ans.)$

A.C. Load Lines

Very often the load into which a transistor or a FET works is not the same for both a.c. and d.c. conditions. When this is the case two load lines must be drawn on the characteristics: a d.c. load line to determine the operating point, and an a.c. load line to determine the current or voltage gain of the circuit. The a.c. load line *must* pass through the operating point.

Fig. 6.20 shows the circuit of a single-stage common-emitter amplifier using potential-divider bias. The emitter decoupling capacitor C_3 has a very high reactance at very low frequencies and does not shunt the emitter resistance R_4 at zero frequency (direct current). The

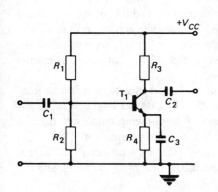

Typical values
$R_1 = 56 \text{ k}\Omega$, $R_2 = 10 \text{ k}\Omega$
$R_3 = 3 \text{ k}\Omega$. $R_4 = 1 \text{ k}\Omega$
$C_1 = C_2 = 10 \, \mu\text{F}$
$C_3 = 47 \, \mu\text{F}$

Fig. 6.20 A single-stage common-emitter amplifier

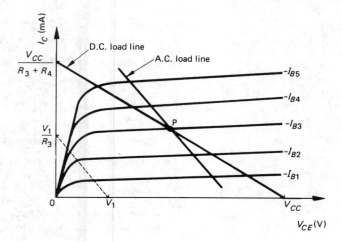

Fig. 6.21 A.C. and d.c. load lines

d.c. load on the transistor is therefore $R_3 + R_4$ ohms. At signal frequencies the reactance of C_3 is low and the a.c. load on the transistor is reduced to R_3 ohms. A d.c. load line is first drawn on the output characteristics between the points

$$I_C = 0, \quad V_{CE} = V_{CC}, \quad \text{and}$$
$$V_{CE} = 0, \quad I_C = V_{CC}/(R_3 + R_4)$$

as shown by Fig. 6.21. A suitable operating point P is then selected.

The a.c. load line must be drawn passing through the operating point with a slope equal to the reciprocal of the a.c. load resistance on the transistor, i.e. a slope of $-1/R_3$. To avoid extending the current axis, proceed as follows. (a) Assume that the a.c. load is actually a d.c. load, and using any convenient value of supply voltage (V_1 in Fig. 6.21), draw lightly the corresponding d.c. load line using the method previously explained, i.e. draw the line between the points V_1 and V_1/R_3. (b) This load line has the required slope, so now draw the actual a.c. load line parallel to it and passing through the operating point.

Another instance where the a.c. load on a transistor is different from the d.c. load is shown by Fig. 6.22. The external load resistance R_L, which may be the input resistance of a following amplifier stage, is coupled to the transistor by the coupling capacitor C_2. If the decoupling capacitor C_3 has negligible reactance at all signal frequencies the d.c. load on the transistor is $R_3 + R_4$ ohms and the a.c. load is R_3 in parallel with R_L, i.e.

$$R_{L(\text{eff})} = R_3 R_L/(R_3 + R_L)$$

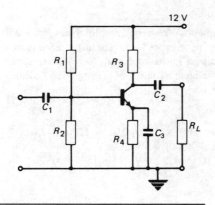

Values
$R_1 = 30\,\text{k}\Omega, \ R_2 = 6.8\,\text{k}\Omega$
$R_3 = 2.5\,\text{k}\Omega, \ R_4 = 0.5\,\text{k}\Omega$
$R_L = 2.5\,\text{k}\Omega, \ C_3 = 47\,\mu\text{F}$
$C_1 = C_2 = 10\,\mu\text{F}$

Fig. 6.22

Example 6.10

The bipolar transistor used in the circuit of Fig. 6.22 has the data given in Table 6.3. Plot the output characteristics of the transistor, draw the d.c. load line and select a suitable operating point. Draw the a.c. load line and use

Table 6.3

V_{CE}(V)	I_C(mA)			
	V_{BE} = 610 mV	V_{BE} = 620 mV	V_{BE} = 630 mV	V_{BE} = 640 mV
3	0.85	1.55	2.32	3.08
5	1.00	1.74	2.56	3.35
7	1.13	1.92	2.76	3.60
9	1.30	2.13	3.00	

it to find the a.c. voltage across the 2500 Ω load R_L when an input signal V_{be} of peak value 15 mV is applied to the circuit. Assume all the capacitors have zero reactance at signal frequencies.

Solution

The output characteristics are shown plotted in Fig. 6.23. The d.c. load on the transistor is $R_3 + R_4 = 3000$ Ω; the d.c. load line must therefore be drawn between the points

$$I_C = 0, \quad V_{CE} = 12 \text{ V} \quad \text{and}$$
$$V_{CE} = 0, \quad I_C = 12/3000 = 4 \text{ mA}$$

Since the input signal has a peak value of 15 mV, a suitable base bias voltage is 625 mV, the operating point is then P.

The a.c. load on the transistor is the 2500 Ω collector resistor in parallel with the 2500 Ω load, i.e. 1250 Ω. An a.c. load line with a slope of −1/1250

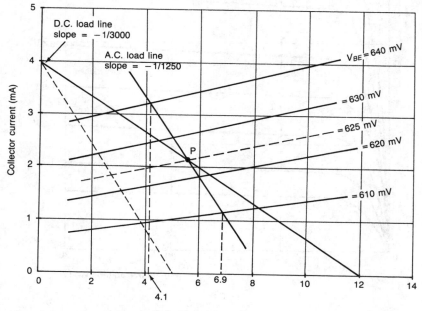

Fig. 6.23

must therefore be drawn on the output characteristics. To draw a d.c. load line with the same slope assume a convenient supply voltage, say 5 V; then this d.c. load line joins the points

$$V_{CE} = 5 \text{ V} \quad \text{and} \quad I_C = 5/1250 = 4 \text{ mA}$$

The equivalent d.c. load line is shown dotted and the wanted a.c. load line has been drawn parallel to it and passing through the operating point.

When a signal of ± 15 mV (peak) is applied to the transistor V_{BE} varies between 610 mV and 640 mV. Projection from the intersection of the load line and the 610 mV and 640 mV V_{BE} curves gives the resulting values of collector voltage as 4.1 V and 6.9 V. The peak collector signal voltage is therefore (6.9 – 4.1)/2 or 1.4 V.

Voltage Gain of FET Amplifier

The voltage gain of a FET amplifier can also be found with the aid of a load line. For example, Fig. 6.24 shows an a.c. load line drawn on the drain characteristics of a FET, and the dotted projections from the load line show how the drain voltage swing, resulting from the application of an input signal voltage, can be found. The voltage gain A_v of the FET amplifier is

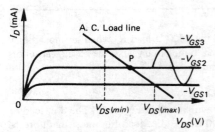

Fig. 6.24 Use of a.c. load line to calculate the voltage gain of a FET amplifier

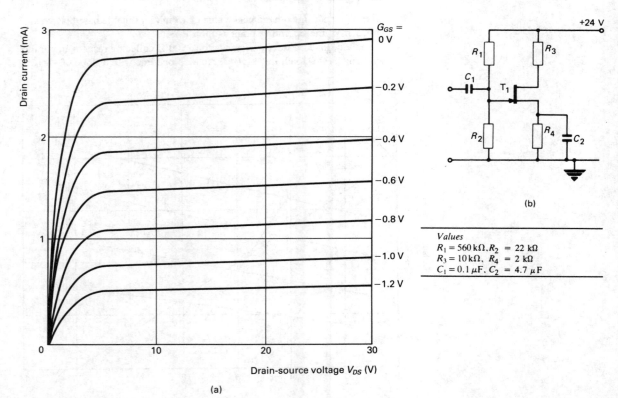

(a)

Fig. 6.25

Values
$R_1 = 560 \text{ k}\Omega, R_2 = 22 \text{ k}\Omega$
$R_3 = 10 \text{ k}\Omega, \ R_4 = 2 \text{ k}\Omega$
$C_1 = 0.1 \ \mu\text{F}, C_2 = 4.7 \ \mu\text{F}$

(b)

$$A_v = \frac{\text{Peak-to-peak drain voltage}}{\text{Peak-to-peak gate-source voltage}} \tag{6.9}$$

$$= \frac{V_{DS(max)} - V_{DS(min)}}{V_{GS3} - V_{GS1}} \tag{6.10}$$

Example 6.11

Fig. 6.25(*a*) shows the drain characteristics of a common-source n-channel JFET, which is used in the single-stage amplifier circuit shown in Fig. 6.25(*b*). Draw the d.c. load line and select a suitable operating point. Draw the a.c. load line and use it to find the voltage gain when a sinusoidal input signal of 0.3 V peak is applied.

Solution

The d.c. load on the FET is 12 kΩ, and hence the d.c. load line must join the points

$$I_D = 0, \quad V_{DS} = 24 \text{ V} \quad \text{and}$$
$$V_{DS} = 0, \quad I_D = V_{DD}/(R_3 + R_4) = 24/(12 \times 10^3) = 2 \text{ mA}$$

The load line has been drawn in Fig. 6.26. A suitable operating point is $V_{GS} = -0.9$ V. The a.c. load line must pass through the chosen operating point

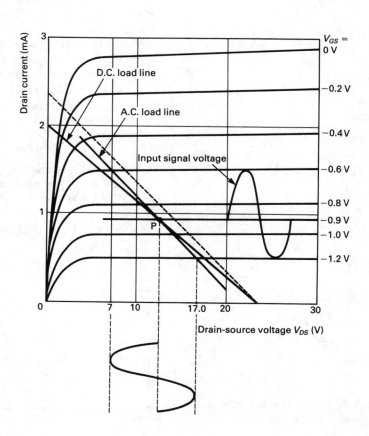

Fig. 6.26

with a slope of $-1/(10 \times 10^3)$ and it has been drawn parallel to the dotted line joining the points

$$I_D = 0, \ V_{DS} = 24 \text{ V} \quad \text{and} \quad V_{DS} = 0, \ I_D = 24/(10 \times 10^3) = 2.4 \text{ mA}.$$

From the a.c. load line, the voltage gain of the circuit is

$$A_v = (17 - 7)/(-1.2 - (-0.6)) = -16.7 \quad (Ans.)$$

Equivalent Circuits

The voltage gain of a transistor amplifier can be calculated using an *equivalent circuit*, or *model*, of the transistor. An equivalent circuit is one that behaves in exactly the same way as the device it represents. Two equivalent circuits are often employed for audio-frequency amplifier calculations; these are the *h parameter* circuit and the *mutual conductance* circuit.

h Parameter Circuit

The *h* parameter equivalent circuit of a bipolar transistor is shown by Fig. 6.27. The three *h* parameters have been met previously in Chapter 3. h_{ie} is the input resistance with V_{CE} constant, h_{fe} is the current gain with V_{CE} constant, and h_{oe} is the output conductance with I_B constant.

When a collector load resistance R_L is connected across the output of the equivalent circuit it will appear in parallel with the output resistance $1/h_{oe}$ of the transistor to give an effective load resistance $R_{L(\text{eff})}$ of $(R_L \times 1/h_{oe})/(R_L + 1/h_{oe})$. The collector current $I_C = h_{fe}I_b$ flows in the total resistance to produce the output voltage $V_{out} = V_{ce}$. Therefore

$$V_{out} = h_{fe}I_b \times R_{L(\text{eff})}$$

Very often $1/h_{oe}$ is much larger than R_L so that $R_{Lo(eff)} = R_L$ which means that h_{oe} can be neglected. The input voltage is $V_{in} = V_{be} = I_b h_{ie}$ so that the voltage gain A_v is

$$A_v = V_{out}/V_{in} = (h_{fe}I_b R_L)/(I_b h_{ie}) = (h_{fe}R_L)/h_{ie} \quad (6.11)$$

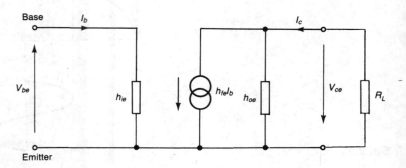

Fig. 6.27 *h* parameter equivalent circuit of a bipolar transistor

Alternatively, the presence of the collector load resistance reduces the current gain of the transistor to a value less than h_{fe}. Applying Kirchhoff's law to the collector circuit:

$$I_c = h_{fe}I_b + h_{oe}V_{ce} \qquad (6.12)$$

An increase in I_c produces an increased voltage drop across the collector resistance and so V_{ce} falls. Therefore, $V_{ce} = -I_cR_L$. Substituting into equation (6.12) gives

$$I_c = h_{fe}I_b - h_{oe}I_cR_L$$
$$I_c(1 + h_{oe}R_L) = h_{fe}I_b$$
$$I_c = h_{fe}I_b/(1 + h_{oe}R_L)$$

Therefore, current gain $A_i = I_c/I_b = h_{fe}/(1 + h_{oe}R_L)$ (6.13)

Example 6.12

A bipolar transistor has the following h parameters: $h_{ie} = 1200\ \Omega$, $h_{fe} = 150$, and $h_{oe} = 60 \times 10^{-6}$ S. The transistor is used as an amplifier with a collector load resistance of 2000 Ω. Calculate the voltage gain of the circuit (a) taking h_{oe} into account, and (b) neglecting h_{oe}.

Solution
  (a) Output resistance $= 1/h_{oe} = 1/(60 \times 10^{-6}) = 16.67$ kΩ
Effective load resistance = 16.67 kΩ in parallel with 2 kΩ, i.e. 1786 Ω
Voltage gain $= (150 \times 1786)/1200 = 223$ (Ans.)
  (b) Alternatively, $A_i = 150/(1 + 60 \times 10^{-6} \times 2000) = 134$
Voltage gain $= (134 \times 2000)/1200 = 223$ (Ans.)
(b) Voltage gain $= (150 \times 2000)/1200 = 250$ (Ans.)

The h parameters of a bipolar transistor are not constant quantities but instead they vary with both the collector current and the temperature. Figs. 6.28(a) and (b) show typical variations of the three h parameters.

Mutual Conductance Circuit

The ratio of h_{fe} to h_{ie}, which is the mutual conductance g_m, is however, much more constant since h_{fe} and h_{ie} vary in different directions. The value of g_m is primarily determined by the d.c. collector current ($g_m = 38$ mS per mA of collector current) and it is easily predictable. Calculations of amplifier gain are therefore often based on the mutual conductance model of the transistor shown by Fig. 6.29. The output resistance r_{out} is often omitted since its effect on the voltage gain is generally small.

Consider the circuit given in Fig. 6.30. At signal frequencies the reactances of the three capacitors are negligibly small. The collector supply line is effectively at earth potential as far as a.c. signals are concerned and so the bias resistors R_1 and R_2 are in parallel both with one another and with the base-emitter terminals of the transistor. The

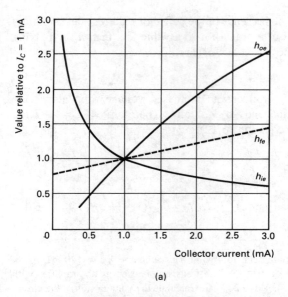

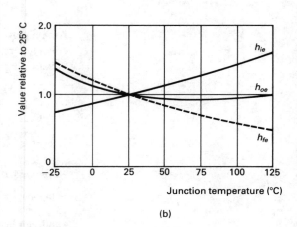

(a)

(b)

Fig. 6.28 Variation of h parameters with (a) collector current and (b) temperature

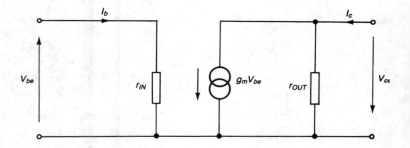

Fig. 6.29 Mutual conductance equivalent circuit of a bipolar transistor

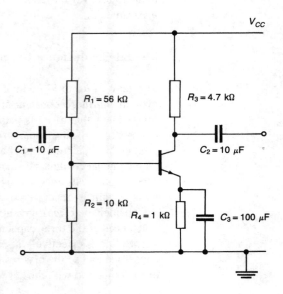

Fig. 6.30

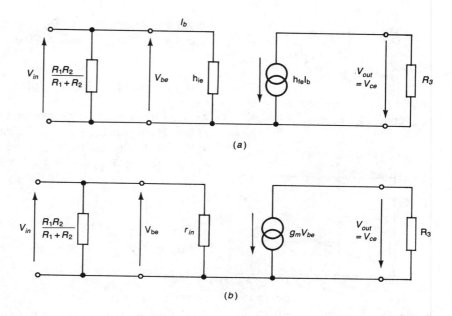

Fig. 6.31 Equivalent circuits of Fig. 6.30 (*a*) *h* parameter and (*b*) mutual conductance

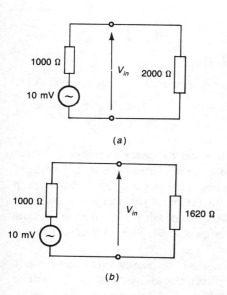

Fig. 6.32

emitter resistor R_4 is effectively short circuited by capacitor C_3 and the collector resistor appears between the collector and the emitter. Thus, the *h* parameter and mutual conductance equivalent circuits of the amplifier are as given in Fig. 6.31(*a*) and (*b*).

The bias resistors shunt the signal path and so they reduce the input resistance of the circuit. To limit this shunting effect the bias resistors should have as high a resistance as possible.

If the parameters of the transistor are $h_{ie} = 2000~\Omega$ and $h_{fe} = 120$ and $g_m = 120/2000 = 60$ mS, then the voltage gain of the circuit is

$$A_v = V_{out}/V_{in} = 120 \times 4700/2000 = 282$$

The presence of the bias resistors does not affect the voltage gain of the circuit but it does reduce the input voltage that appears across the input terminals. Suppose that the source voltage has an e.m.f. of 10 mV and an internal resistance of 1000 Ω. Then, Fig. 6.32(*a*), the input voltage is $V_{in} = 6.67$ mV, $V_{out} = 1.88$ V, if the bias resistors are neglected. If, Fig. 6.32(*b*), the bias resistors are taken into account, the input resistance of the amplifier r_{in} is found from $1/r_{in} = 1/2000 + 1/(56 \times 10^3) + 1/(10 \times 10^3)$ and $r_{in} = 1620~\Omega$. Now the input voltage is 6.18 mV and $V_{out} = 1.74$ V.

The difference between the two output voltages obtained may seem to be fairly significant but it must be remembered that the quoted resistor values are all nominal values and are subject to tolerance

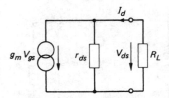

Fig. 6.33 FET equivalent circuit

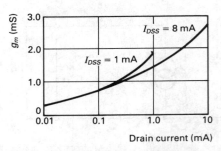

Fig. 6.34 Variation of FET mutual conductance with drain current

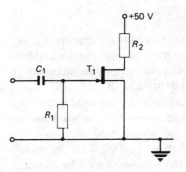

Fig. 6.35

Design of a Single-stage Audio-frequency Amplifier

variations. Because of this, for practical purposes g_m is often taken as being equal to $I_C/25$ mS, or 40 mS per mA of collector current.

Voltage Gain of a FET Amplifier

At audio-frequencies the performance of a FET can be described by its mutual conductance $g_m = I_d/V_{gs}$, and the equivalent circuit shown in Fig. 6.33. The resistance r_{ds} is the output resistance of the FET and is equal to $\delta V_{DS}/\delta I_D = V_{ds}/I_d$ with V_{GS} constant. The output voltage V_{ds} of the circuit is given by

$$V_{ds} = (g_m V_{gs} R_L r_{ds}/(R_L + r_{ds})$$

Therefore the voltage gain A_v is given by

$$A_v = V_{ds}/V_{gs} = g_m R_L r_{ds}/(r_{ds} + R_L) \tag{6.14}$$

If, as is usual, $r_{ds} \gg R_L$,

$$A_v = g_m R_L \tag{6.15}$$

The mutual conductance of a FET is not a constant quantity but instead is a function of the drain current as shown in Fig. 6.34. (I_{DSS} is the drain current for $V_{GS} = 0$.)

Example 6.13

Calculate the drain load resistance required to give the circuit of Fig. 6.35 a voltage gain of 20. The FET used has $g_m = 4 \times 10^{-3}$ S and $r_{ds} = 100$ kΩ.

Solution

From equation (6.14), $A_v = 20 = (4 \times 10^{-3} \times 10^5 R_2)/(10^5 + R_2)$

Therefore $R_2 = (20 \times 10^5)/380 = 5.26$ kΩ *(Ans.)*

Alternatively, using the approximate expression given by equation (6.15)

$$A_v = 20 = 4 \times 10^{-3} R_2 \quad \text{or} \quad R_2 = 20/(4 \times 10^{-3}) = 5000 \ \Omega \quad (Ans.)$$

In the design of a single-stage audio-frequency amplifier a number of factors must be taken into account. These include the choice of operating point and the required voltage gain. Other factors such as the required bandwidth and the noise performance are also of importance but are beyond the scope of this book. There are several different approaches to the design of an amplifier available and here just one of them is given.

Suppose that an amplifier stage is to deliver the maximum possible output voltage so that the d.c. collector-emitter voltage V_{CE} will be set at one-half of the supply voltage V_{CC}. Assuming for the moment, that the rest of the supply voltage is dropped across the collector resistor R_3, then $V_{CC}/2 = I_C R_3$. This gives the maximum possible output voltage. The voltage gain $A_v = g_m R_3 = 40 I_C \times V_{CC}/2 I_C = 20 V_{CC}$. This approximate relationship will allow a choice to be made of the collector supply voltage to allow a wanted gain to be obtained, or alternatively, if the supply voltage is already fixed by other considerations, it will establish the maximum possible voltage gain.

A suitable d.c. load line can be drawn on the output characteristics of the transistor and the operating point located at $V_{CC}/2$ volts. This allows the collector current to be determined. Alternatively, a collector current may be arbitrarily selected, which will very often be the value quoted in the data sheet for the typical value of h_{FE}.

The collector resistor can then be determined from $R_3 = V_{CC}/2 I_C$.

The potential divider bias circuit must establish the required operating point and also provide adequate d.c. stability. If $h_{FE} R_4 \gg R_1 R_2/(R_1 + R_2)$ the collector current will be very nearly equal to $(V_B - V_{BE})/R_4$, i.e. it will be independent of h_{FE}. The collector current is then still subject to variations due to changes in V_{BE} but this factor may be minimized by making V_B several times larger than V_{BE}. Since $V_E = V_E + V_{BE}$ this means that the voltage across R_4 should not be less than about 1 V. A good rule is to make V_E equal to $V_{CC}/10$ with a minimum value of 1 V.

For the circuit to operate correctly the base voltage V_B should remain more or less constant at $V_B = V_{CC} R_2/(R_1 + R_2)$ as the collector current and thus $V_E = I_C R_4$ varies. To achieve this the current flowing through the lower bias resistor R_2 must be several times larger, by a factor n, than the base current $I_B = I_C/h_{FE}$. The larger the factor n the better will be the d.c. stability of the circuit but the lower will be the values of the resistors R_1 and R_2. Since R_1 and R_2 are effectively in parallel, both with the signal path and with one another, their values must not be too small. This means that the choice of n must be a compromise between the a.c. and d.c. performances of the circuit. A suitable choice for n is 10.

Then, $R_2 = V_B/n I_B$ and $R_1 = (V_{CC} - V_B)/(n+1) I_B$. Calculation of suitable values for the coupling and decoupling capacitors is more difficult and values such as those quoted in Table 6.1 should be used.

Example 6.14

Design an audio-frequency single-stage amplifier to have a voltage gain of about 240 using the transistor whose characteristics are given in Fig. 6.36. The maximum possible output voltage is required.

Solution

Voltage gain $= 240 = 20 V_{CC}$, or $V_{CC} = 240/20 = 12$ V. The operating point is then set at $V_{CE} = 6$ V, $I_B = 15\ \mu$A. The d.c. load line has been drawn through these two points. The d.c. collector current is then 1.5 mA and $h_{FE} = (1.5 \times 10^{-3})/(15 \times 10^{-6}) = 100$.

The slope of the load line is $(12 - 6)/(1.5 - 0) \times 10^{-3} = 4000\ \Omega$. The emitter voltage $V_E = V_{CC}/10 = 12/10 = 1.2$ V. Hence, $R_4 = 1.2/(1.5 \times$

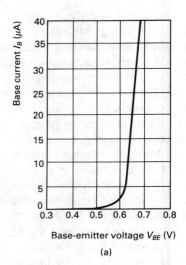

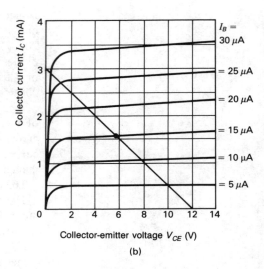

Fig. 6.36

(a)   (b)

$10^{-3}) = 800\ \Omega$. Choosing the nearest preferred values for R_3 and R_4; $R_3 = 3900\ \Omega$ and $R_4 = 820\ \Omega$. Now the voltage drop across $(R_3 + R_4)$ is $1.5 \times 10^{-3} \times (3900 + 820) = 7$ V which leaves V_{CE} as 5 V.

The base current $I_B = I_C/h_{FE} = 15\ \mu$A. Therefore, choose the current in R_2 to be 150 μA. From Fig. 6.36a, when $I_B = 15\ \mu$A, $V_{BE} = 0.65$ V, giving $V_B = 1.2 + 0.65 = 1.85$ V. Then,

$$R_2 = 1.85/(10 \times 15 \times 10^{-6}) = 12.3\ \text{k}\Omega$$

and

$$R_1 = (12 - 1.85)/(11 \times 15 \times 10^{-6}) = 61.5\ \text{k}\Omega$$

Choosing the nearest preferred values, $R_2 = 12$ kΩ and $R_1 = 62$ kΩ.

The midband voltage gain will be $g_m R_3 = 38 \times 1.5 \times 10^{-3} \times 3900 = 222$. To increase the gain nearer to the wanted target of 240 the next higher preferred value for R_3 could be used, i.e. $R_3 = 4300\ \Omega$. Then $A_v = 245$ but the maximum output voltage will be reduced.

Multi-stage Amplifiers

Very often the voltage gain that an amplifier is required to provide is greater than can be supplied by a single stage. It is then necessary to connect two, or more, stages in cascade to obtain the required gain. The term 'cascade' means that the output terminals of one stage are connected to the input terminals of the next stage. The overall voltage gain of a number of cascaded amplifier stages is equal to the product of their individual gains, i.e. $G = G_1 \times G_2 \times G_3 \ldots$

Example 6.15

An amplifier stage has a voltage gain of A_v. If two such stages are connected in cascade to obtain an overall voltage gain of 1000, what is the required gain per stage?

Solution

$$G_1 \times G_2 = G^2 = 1000$$
$$\sqrt{1000} = 31.62 \quad (Ans.)$$

Exercises

Fig. 6.37

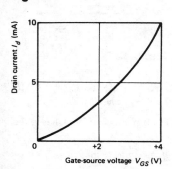

Fig. 6.38

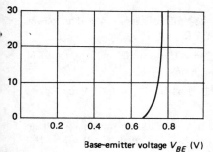

Ig. 6.39

6.1 Fig. 6.37 shows how the d.c. current gain h_{FE} of a transistor varies with the collector current I_C. If the maximum gain is required determine the value of I_C that should be chosen.

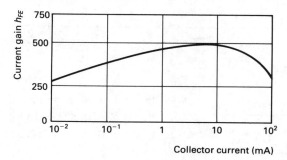

6.2 List the factors that should be taken into account when choosing a suitable operating point for a transistor.

6.3 The mutual characteristic of an n-channel enhancement-mode MOSFET is shown in Fig. 6.38. The bias voltage is + 2 V. If a sinusoidal signal of peak value 1 V is applied to the device determine the waveform of the drain current.

6.4 Explain, with the aid of a typical mutual characteristic, why Class B or Class C bias cannot be used with a single-ended resistance-loaded amplifier.

6.5 Fig. 6.39 shows a typical mutual characteristic for a bipolar transistor. Determine suitable base-emitter bias voltages for (*a*) Class A, (*b*) Class B and (*c*) Class C bias.

6.6 Explain why d.c. stabilization of a FET is necessary even though thermal runaway is not a problem. Hence, explain the disadvantage of the circuit given by Fig. 6.40(*a*).

6.7 In Fig. 6.41(*a*), $V_{DD} = 12$ V, $R_3 = 3$ kΩ and $V_{DS} = 6$ V. Calculate the drain current. If $R_1 = R_2$ calculate V_{GS} also.

6.8 In the circuit of Fig. 6.40(*b*), $R_1 = 470$ kΩ, $R_2 = 3$ kΩ, $R_3 = 1.2$ kΩ and $V_{DD} = 12$ V. Calculate the gate-source bias voltage if the drain current is 1.5 mA.

Fig. 6.40

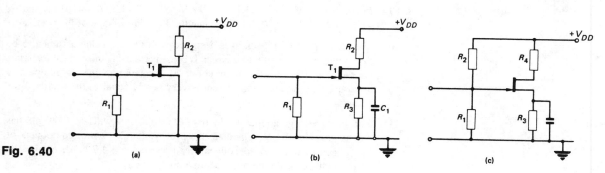

(a)　　　　　(b)　　　　　(c)

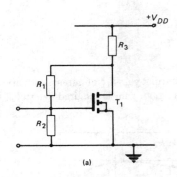

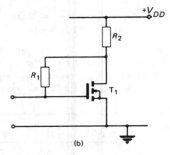

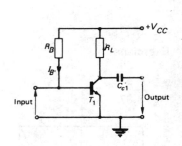

Fig. 6.41   Fig. 6.42

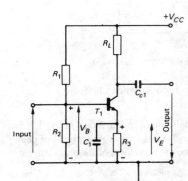

Fig. 6.43

Fig. 6.44

6.9 In Fig. 6.40(c), $R_2 = 150$ kΩ, $R_1 = 22$ kΩ, $R_3 = 2$ kΩ and $R_4 = 3.8$ kΩ. If $V_{DD} = 15$ V and $I_D = 1.5$ mA calculate (a) V_{DS} and (b) V_{GS}.

6.10 In Fig. 6.41(b) $R_1 = 300$ kΩ and $R_2 = 4.7$ kΩ, if $V_{DS} = 5$ V, what is the value of V_{GS}?

6.11 In the circuit of Fig. 6.42, $V_{CC} = 6$ V and $R_B = 470$ kΩ. If $V_{BE} = 0.61$ V, calculate the base current I_B.

6.12 In Fig. 6.43 the collector-to-earth voltage is 9 V, $R_L = 3$ kΩ and $I_C = 2$ mA. Calculate the supply voltage V_{CC}. If the emitter resistor is 1 kΩ calculate V_E.

6.13 In Fig. 6.43 $V_{CC} = 12$ V, the collector-to-earth voltage is 6 V and the power dissipated in R_L is 6 mW. Calculate (a) R_L and (b) I_C.

6.14 In Fig. 6.43 $R_3 = 1.2$ kΩ and $I_C = 1.8$ mA. If $h_{FE} = 100$ and $V_{BE} = 0.62$ V calculate (a) I_B and (b) the voltage across R_2.

6.15 The transistor used in the circuit of Fig. 6.44 has $V_{BE} = 0.63$ V and $h_{FE} = 85$. Calculate I_C.

6.16 In Fig. 6.43 $V_{CC} = 12$ V and $I_C = 2$ mA. If 1/10th of the supply voltage appears across R_3 calculate R_3. If $V_{CE} = V_{CC}/2$ calculate R_L. If $h_{FE} = 100$ calculate I_B. If $I_{R2} = 10I_B$ calculate R_1 and R_2. $V_{BE} = 0.65$ V.

6.17 For the circuit of 6.20, calculate the d.c. input power. Also find the collector dissipation for zero signal conditions.

6.18 A transistor has $h_{FE} = 120$ and $I_{CBO} = 20$ nA. Calculate its collector current when the base current is 20 μA.

6.19 In the circuit of Fig. 6.43 $V_{CC} = 12$ V, $R_1 = 33$ kΩ, $R_2 = 10$ kΩ and $R_3 = 1.2$ kΩ. If $I_C = 1.75$ mA calculate the V_{BE} bias voltage of the transistor.

6.20 The bias circuit of Fig. 6.20 is designed with the h_{FE} value assumed to be the nominal value of 100. One circuit is constructed using this type of transistor where the h_{FE} value is the minimum of 70. Explain how the circuit operates to ensure that the collector current is very nearly equal to the designed-for value.

6.21 The signal voltage applied to the base of a transistor with $g_m = 40$ mS has a peak value of 0.1 V. Calculate the peak value of the a.c. component of the collector current. If the collector load resistance is 4.7 kΩ determine the output voltage of the circuit and the voltage gain.

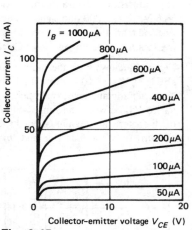

Fig. 6.45

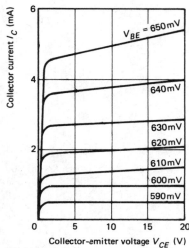

Fig. 6.46

6.22 An audio-frequency amplifier uses a FET with $r_{ds} = 10$ kΩ and $g_m = 5$ mS. What value of drain load resistor is needed to give a voltage gain of 40?

6.23 On the output characteristics given in Fig. 6.45 draw the load line for a d.c. load of 400 ohms. The operating point is $I_B = 150$ μA and the supply voltage is 20 V. If the emitter resistance is 100 ohms calculate V_E, V_{CE} and the collector-to-earth voltage. Determine the peak-to-peak output voltage when a sinusoidal voltage varies the base current by ± 50 μA.

6.24 Fig. 6.46 shows the collector current of a transistor plotted against collector-emitter voltage for various values of base-emitter voltage. The transistor is used in an amplifier (Fig. 6.43) with $V_{CC} = 20$ V, $R_L = 3.5$ kΩ and $R_3 = 500$ Ω. Draw the load line and choose a suitable operating point to give the maximum possible output voltage. Determine the peak-to-peak collector current when the peak base voltage change is 25 mV. Calculate the voltage gain of the circuit (a) from the load line, and (b) using the expression $A_v = g_m R_L$.

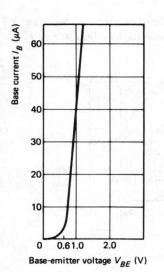

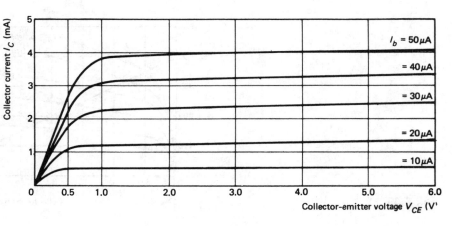

Fig. 6.47   (a)   (b)

6.25   Figs 6.47(a) and (b) show, respectively, the input and output characteristics of a transistor. The transistor is used in a single-stage amplifier with V_{CC} = 6 V, R_L = 1800 ohms and emitter resistance is 200 ohms. Draw the load line and mark the operating point for a base bias current of 20 μA. Determine the required bias voltage V_{BE}. Calculate the ratio $\delta I_C / \delta V_{BE}$.

6.26   Fig. 6.48 shows both the mutual and drain characteristics of an n-channel JFET. If V_{DD} = 20 V draw the load line for R_L = 2000 ohms on the drain characteristic and select the operating point V_{GS} = – 2 V. A signal voltage varies V_{GS} between the limits – 1 V and – 3 V. Determine from both sets of characteristics the mutual conductance of the device. Calculate the voltage gain (a) from the load line and (b) using the expression $A_v = g_m R_L$.

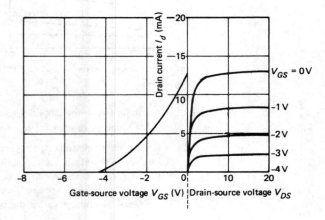

Fig. 6.48

6.27   Fig. 6.49 shows a transistor output characteristic with a load line drawn on it. Determine the load to which this relates. Caculate the maximum peak-to-peak output voltage and current if the operating point is (a) I_B = 1 mA, (b) I_B = 3 mA and (c) I_B = 5 mA. Comment on the results.

6.28   For Fig. 6.49 calculate the d.c. power taken from the supply and the collector dissipation under no-signal conditions. Assume the base-bias current I_B to be 3 mA.

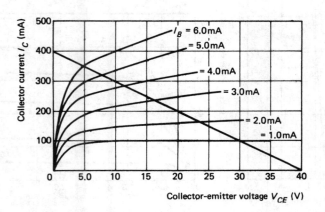

Fig. 6.49

7 Power Supplies

All electronic equipments require a power supply of some kind to provide the necessary d.c. operating voltages and currents. Some portable equipments, such as pocket calculators and small radio receivers, are battery operated but most equipments employ an electronic power supply. A linear power supply converts an a.c. voltage (usually the public mains supply) into a d.c. voltage supply. It consists of: (*a*) a transformer whose function is to convert the a.c. mains supply voltage to the lower value required by the equipment, and also to provide isolation between the a.c. supply's neutral line and the power supply's common line; (*b*) a rectifier unit whose function is to convert the a.c. voltage supplied by the transformer to a d.c. voltage; (*c*) a filter whose purpose is to remove *ripple* from the rectified voltage; and (*d*) (usually) a voltage regulator to keep the d.c. output voltage more or less constant. The regulator may be a discrete circuit or an IC device.

The correct operation of many equipments demands that the d.c. power supply voltage is maintained at a constant value, within fairly fine limits, even though the input mains voltage and/or the current taken from the power supply may vary. Generally, the inherent regulation of a supply is inadequate to meet the demands placed on it by the supplied equipment and then some kind of voltage regulator circuitry must be provided. The function of a voltage regulator is to maintain a constant voltage across the load as the input voltage and/or the load current vary within specified limits. The block diagram of a regulated power supply is shown by Fig. 7.1.

The main requirements of a power supply are:

(*a*) to provide the rated voltage for the circuit to be supplied with a specified ± tolerance
(*b*) to be able to supply the rated maximum current to the supplied

Fig. 7.1 Block diagram of a regulated power supply

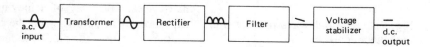

circuit without the supply voltage falling outside specified limits

(c) to maintain the supplied voltage constant within specified limits as the load changes, or the mains supply input voltage varies, or the ambient temperature changes. This requirement is known as the *regulation*.

A specification for a power supply might include:

(a) input voltage 240 V a.c.
(b) output voltage ± 15 V ± 2%
(c) maximum output current 1 A
(d) regulation 3% (see page 152).

Rectifier Circuits

A number of different circuits exist that are capable of converting an a.c. supply into a pulsating d.c. current and they may be broadly divided into either half-wave rectifiers or full-wave rectifiers.

Half-wave Rectification

In its simplest form, half-wave rectification consists merely of the connection of a diode in series with the a.c. supply and the load, as shown in Fig. 7.2(a).

The diode conducts only during those alternate half-cycles of the a.c. secondary voltage V_s that make point A positive relative to point B, and so the load current consists of a series of half sine-wave pulses. The voltage V_{DC} developed across the load is the product of the load current and the load resistance and it has the same waveform as the load current (Fig. 7.2(b)). The mean, or average, value of the load voltage is equal to $V_s/2$ volts. V_s is normally quoted as an r.m.s. value.

The disadvantage of this simple rectifying circuit is very clear: the load voltage, although unidirectional, varies considerably and is, indeed, zero for half of the time. Its value is $V_{DC} = 0.45\ V_s$. Such a waveform is only suitable for simple applications, such as battery charging, since the voltage variations will appear as noise at the output of any equipment operated from the supply. When the diode is non-conducting, the peak voltage across it, known as the *peak inverse voltage* (PIV), is equal to the peak value of the transformer secondary voltage. This voltage must not exceed the peak reverse repetitive voltage rating of the type of diode employed.

The d.c. output of a rectifier circuit is required to be as steady as possible and a great step towards this goal would be achieved if the load voltage could be prevented from falling to zero during alternate half-cycles. One way of achieving this is to connect a reservoir capacitor C_1 in parallel with the load resistance as shown in Fig. 7.3(a). Each time the diode conducts, the current that flows charges

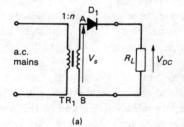

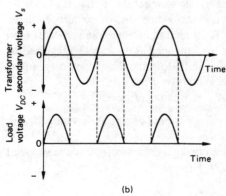

Fig. 7.2 The half-wave rectifier with resistance load

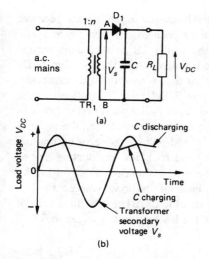

Fig. 7.3 The half-wave rectifier with resistance-capacitance load

Typical values

$R_L = 20 \ \Omega$ $C \ 1000 \ \mu F$

Fig. 7.4 Showing the effect of load changes on a half-wave rectifier with resistance-capacitance load

the capacitor and the voltage across the capacitor builds up. During the intervals of time when the diode is non-conducting, the capacitor discharges via the load resistance and prevents the load voltage falling to zero (Fig. 7.3(*b*)). The capacitor continues to discharge, at a rate determined by the time constant C_1R_L seconds, until the point A is taken more positive than the capacitor voltage by a positive half-cycle of the voltage V_s. The diode then conducts, the capacitor is recharged and the capacitor voltage rises again. If the load current is fairly small the capacitor does not discharge very much between charging pulses and the average load voltage V_{DC} is only slightly less than the peak value of the secondary voltage V_s. $V_{DC} = 1.41 \ V_s$. The PIV is increased to twice the peak secondary voltage V_s because of the polarity of the voltage developed across C.

The value of the load voltage is adjustable, within limits, by suitable choice of the turns ratio n of the input transformer. An increase in the load, i.e. in the load current, means that the load resistance has been reduced; this, in turn, means that the time constant of the discharge path is smaller. Capacitor C_1 then discharges more rapidly and the load voltage is not as constant (see Fig. 7.4).

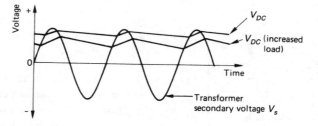

A completely steady load voltage cannot be obtained in this way since too large a value of capacitance would be required. The maximum value of capacitance that can be employed is limited, because the larger the capacitance value the greater is the current required to charge the capacitor to a given voltage, and the maximum current that can be handled by a diode is limited to a figure quoted by the manufacturer. The fluctuating, unidirectional voltage appearing across the load may be regarded as a d.c. voltage having an a.c. voltage superimposed on it. This a.c. voltage is known as the *ripple voltage* and it is at the frequency of the supply voltage, usually 50 Hz. The ripple voltage is undesirable, since the object of rectification is to provide a steady d.c. voltage, and can be removed by a smoothing, or filter, circuit connected between the diode and the load.

Full-wave Rectification

With full-wave rectification of an a.c. source, both half-cycles of the input waveform are utilized and alternate half-cycles are inverted to

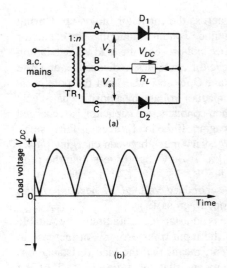

(a)

(b)

Fig. 7.5 The full-wave rectifier with resistance load

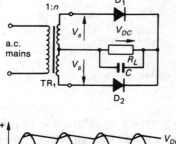

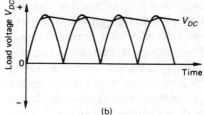

(b)

Fig. 7.6 The full-wave rectifier with resistance-capacitance load

Typical values

$R_L = 50 \ \Omega$ $C = 2200 \ \mu F$

give a unidirectional load current. The circuit of a full-wave rectifier is shown in Fig. 7.5(a) and can be seen to use two diodes. The secondary winding of the input transformer TR_1 is accurately centre-tapped so that equal voltages are applied across the two diodes D_1 and D_2. During those half-cycles of the input waveform that make point A positive with respect to point B and point C negative relative to point B, diode D_1 conducts and D_2 does not and current flows in the load in the direction indicated by the arrow. When the point C is positive with respect to point B, and point A is negative relative to point B, diode D_2 is conducting and D_1 is non-conducting and so current flows in the load in the same direction as before. The waveform of the current, and hence of the load voltage V_{DC}, is shown in Fig. 7.5(b). The PIV is equal to $2V_s$. The peak load voltage is $\sqrt{2}V_s$ and the average value of the load voltage is equal to 0.637 times $V_{s(max)}$ or $0.9V_s$.

Example 7.1

Calculate the average output voltage of the circuit shown in Fig. 7.5 if the mains voltage is 240 V and the transformer turns ratio n is 4:1.

Solution
$$V_s = 240/4 = 60 \text{ V}$$
$$V_{DC} = 0.9 \times 60 = 54 \text{ V} \quad (Ans.)$$

A more constant value of load voltage can be obtained by the connection of a capacitor across the load as shown in Fig. 7.6(a). The action of the reservoir capacitor is exactly the same as in the half-wave circuit but now the capacitor is re-charged twice per input cycle instead of only once. Between charging pulses the capacitor starts to discharge through the load but, provided the time constant $C_1 R_L$ is not too short, the load voltage has not fallen by much before the next charging pulse occurs (Fig. 7.6(b)). The load voltage attains a mean value only slightly less than the peak voltage appearing across one half of the input transformer secondary winding, i.e. $1.14V_s$. The ripple content of the load voltage increases with increase in load current and it can be reduced by the use of a suitable filter network.

The full-wave circuit has a number of advantages over the half-wave circuit: it is more efficient; little, if any, d.c. magnetization of the transformer core occurs; and the ripple voltage is at twice the supply frequency, i.e. at 100 Hz. The increase in the ripple frequency makes it easier for a filter circuit to reduce the percentage ripple to a desired level. The disadvantages of the circuit are the need for a centre-tapped transformer and for two diodes.

Bridge Rectifier

An alternative method of full-wave rectification is the use of a bridge network (see Fig. 7.7). The bridge rectifier circuit uses four diodes

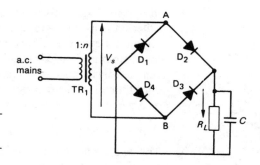

Fig. 7.7 The bridge rectifier

Typical values

$R_L = 30\ \Omega \quad C = 2200\ \mu F$

instead of two but it avoids the need for a centre-tapped input transformer. Further, the arrangement gives a load voltage that is nearly twice as great as that from the circuit of Fig. 7.6(a) — assuming, of course, the same transformer secondary voltage. During those half-cycles of the input that make point A positive with respect to point B, diodes D_2 and D_4 are conducting and diodes D_1 and D_3 are non-conducting; current therefore flows from point A to point B via D_2, the load R_L and D_4. When point A is negative relative to point B, D_1 and D_3 are conducting and D_2 and D_4 are non-conducting; current then flows from point B to point A via D_3, the load R_L and D_1. Both currents pass through the load in the same direction and so a fluctuating, unidirectional voltage is developed across the load having the waveform shown in Fig. 7.5(b) and of mean value 0.9 V_S. The variations in load voltage can be reduced by the connection of a capacitor across the load; the load voltage waveform is then that of Fig. 7.6(b). The d.c. voltage across the load is equal to $1.41V_s$, which is the same as for the circuit of Fig. 7.6.

When higher d.c. voltages are required, the bridge circuit has some advantages over the circuit using a centre-tapped transformer: the PIV of each diode is only equal to the peak secondary voltage V_s; a centre-tapped secondary winding is not required; and the current rating of the transformer is less. This means that a smaller and hence cheaper transformer can be used.

Four-diode bridge rectifier units are available in single packages, for example, the BY 164 series which have $I_{FM} = 1.5$ A and $V_{RMM} = 60$ V.

Filter Circuits

A power supply unit for electronic equipment must provide a d.c. voltage of minimum ripple content. To reduce the ripple voltage to a tolerable level it is generally necessary to include some kind of filter circuit between a rectifier and its load. The simple capacitor connected in shunt across the load reduces ripple and is, therefore, a simple filter that may be adequate for some applications. Further smoothing of the output voltage can be achieved if the capacitor is followed by an LC network, or the rectifier may feed directly into a series inductor. The effectiveness of a filter can be judged in terms of the reduction

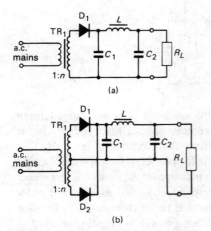

(a)

(b)

Fig. 7.8 The capacitor-input *L-C* filter

Typical values

(a) $C_1 = 32\ \mu F$ $C_2 = 32\ \mu F$ L_1] 30 H
(b) $C_1 = C_2 = 8\ \mu F$ $L = 15$ H

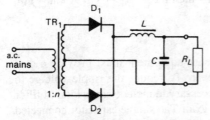

Fig. 7.9 The choke-input filter

Typical value:
$L = 20$ H
$C = 22\ \mu F$

in ripple voltage it gives, but another and (usually) equally important criterion is its voltage regulation. The *load regulation* of a rectifier circuit is a measure of how its output voltage changes as the current taken from it is varied. Ideally, of course, the output voltage should remain constant, so good regulation implies that the voltage changes very little as the load current is varied from minimum to maximum. The percentage regulation of a power supply is given by

$$\% \text{ regulation} = (V_{no\text{-}load} - V_{full\text{-}load})/V_{no\text{-}load} \times 100\% \quad (7.1)$$

The *line regulation* is a measure of the change in output voltage because of a change in the input voltage, with the load current held constant. Typical regulation curves are given in Fig. 7.11.

Example 7.2

A power supply is rated at 15 V, 20 W and has a load regulation of 8%. Calculate its full-load and no-load output currents and voltages.

Solution

Full-load current = 20/15 = 1.33 A   (*Ans.*)
Full-load voltage = 15 V   (*Ans.*)
No-load current = 0 A   (*Ans.*)
No-load voltage = (full-load voltage)/(1 − 8/100) = 15/(1 − 0.08)
= 16.3 V   (*Ans.*)

Example 7.3

A 12 V power supply has a line regulation of 2.5%. Calculate the maximum change in the output voltage if the mains voltage changes to 230 V.

Solution

$2.5 = [(V_{DC} - V_x)/V_{DC}] \times 100$
$2.5V_{DC} = 100V_{DC} - 100V_x$
$100V_x = 97.5V_{DC}$
$V_x = 0.975V_{DC} = 0.975 \times 12 = 11.7$ V

Therefore, change in output voltage = 12 − 11.7 = 0.3 V   (*Ans.*)

Capacitor-input Filters

A capacitor-input filter consists of a shunt capacitor, connected across the output terminals of the rectifier, followed by a basic low-pass filter. The low-pass filter consists of a series inductor and a shunt capacitor as shown by Fig. 7.8. The value of capacitance C_1 is chosen to give a reasonably smooth output voltage from the rectifier proper, and the values of L and C_2 are chosen to give adequate ripple suppression.

The disadvantages of the capacitor-input *LC* filter are (*a*) its cost, weight, size and external fields of the series inductor, and (*b*) its relatively poor load regulation.

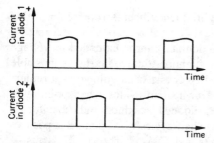

Fig. 7.10 Current waveforms in the choke-input filter

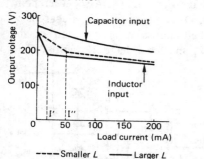

----Smaller *L*   ——Larger *L*

Fig. 7.11 Regulation curves for capacitor- and choke-input filters

Voltage Multiplication

Choke-input Filters

When a choke-input filter is used there is no reservoir capacitor and the rectifier feeds directly into the filter as shown in Fig. 7.9. Inductor *L* and capacitor *C* form a potential divider across the output of the rectifier and reduce the ripple voltage to a low value. The choke-input filter can only be used in conjunction with a full-wave rectifier since it requires current to flow at all times, the current being provided first by D_1 and then by D_2. The current waveforms in the circuit are shown in Fig. 7.10. The fact that current flows continuously, instead of in a series of pulses as in the capacitor-input filter circuits, means that the input transformer is used more efficiently. A further advantage is that the ripple content at the output of the filter is less dependent on the load current. The d.c. load voltage is $0.9V_s$.

Fig. 7.11 shows typical voltage regulation curves for the two types of filter. It can be seen that although the output voltage of the choke-input filter is the smaller, its regulation is better. The regulation can be improved by the use of a larger value of inductance as shown by the two lower curves. If the load current falls below a certain critical value I' or I'' (this could happen each time the equipment was first switched on), the output voltage will rise abruptly. To prevent this happening a resistor, known as a 'bleeder', can be connected across the output terminals of the filter to ensure that a current greater than the critical value is always taken.

As an alternative to the use of a bleeder resistor with its consequent power dissipation, a *swinging choke* can be used. This is an inductor whose inductance depends on the magnitude of the d.c. current flowing in its windings. Typically, such an inductor might have an inductance of 30 H with zero current flow and of only 5 H with 250 mA flowing.

The use of a suitable combination of rectifiers and capacitors can give a d.c. output voltage that is several times greater than the peak voltage appearing across the secondary winding of the input transformer. Consider, for example, Fig. 7.12 which shows a voltage-doubling circuit. During the half-cycles of the input when point A is positive with respect to point B, diode D_1 conducts and capacitor C_1 is charged to the peak voltage V_s appearing across the secondary winding of transformer TR_1. When point B is positive relative to point A, diode D_2 conducts and capacitor C_2 is charged to the same voltage. Capacitors C_1 and C_2 are connected in series across the output terminals and so the voltage appearing across these terminals is equal to twice the peak secondary voltage, i.e. $2V_s$. If large values of capacitance are used and the load current is fairly small, the output voltage has a small ripple content and good regulation. Since the capacitors are charged during alternate half-cycles, the ripple

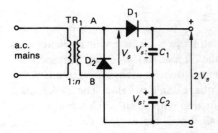

Fig. 7.12 The voltage doubler

frequency is at twice the supply frequency, that is 100 Hz for 50 Hz mains.

The principle of the voltage doubler can be extended to voltage tripling, quadrupling or even higher. Fig. 7.13 shows possible arrangements for (*a*) a voltage tripler and (*b*) a voltage quadrupler. Consider the voltage tripler. During half-cycles when point A is positive with respect to point B, diode D_1 conducts and capacitor C_2 is charged to V_s volts. During the half-cycles when point B is positive relative to point A, D_2 conducts and C_1 is charged to $2V_s$ volts — because the voltage applied across it is the sum of the transformer secondary voltage V_s and the voltage V_s across C_2. Also, when point A is positive relative to point B, D_3 conducts and C_3 is charged to $3V_s$, because the voltage applied across it is the sum of the transformer secondary voltage V_s and the voltage $2V_s$ across C_1.

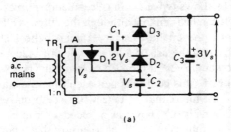

(a)

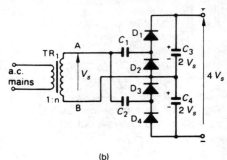

(b)

Fig. 7.13 (*a*) The voltage tripler and (*b*) the voltage quadrupler

Voltage Regulators

For modern electronic equipments a power supply consisting of a transformer, a rectifier and a filter has an inadequate performance. Firstly, the voltage regulation is not good enough and, secondly, the d.c. output voltage varies with change in the a.c. mains voltage. To improve the constancy of the d.c. output voltage as the load and/or the a.c. input voltage vary, a voltage regulator circuit must be employed.

Zener Diode Voltage Regulator

The simplest voltage regulator circuit consists merely of a resistor R_1 connected in series with the input voltage and a Zener diode connected in parallel with the load resistance R_L as shown by Fig. 7.14.

Applying Kirchhoff's voltage law to the circuit,

$$V_{IN} = (I_D + I_L)R_1 + V_{OUT}$$

Rearranging

$$R_1 = (V_{IN} - V_{OUT})/(I_D + I_L) \qquad (7.2)$$

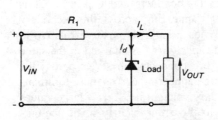

Fig. 7.14 The Zener diode voltage regulator

When a Zener diode is operated in its breakdown region, the current flowing through the diode can vary considerably with very little change in the voltage across the diode. If the load current should increase, the current through the Zener diode will fall by the same percentage in order to maintain a constant voltage drop across R_1 and hence a constant output voltage. Should the load current decrease, the diode will pass an extra current such that the sum of the two currents flowing in R_1 is maintained constant, and the output voltage of the circuit is regulated, or stabilized.

The other cause of output voltage variations is change in the voltage applied across the input terminals of the circuit. If the input voltage should increase, the Zener diode will pass a larger current so that the extra voltage is dropped across resistor R_1. Conversely, if the supply voltage falls, the diode takes a smaller current and the voltage dropped across R_1 is reduced. Because of the varying voltage drop across R_1, the load voltage fluctuates to a much lesser extent than does the input voltage.

Calculation of Series Resistance

(a) Varying load; fixed supply voltage

When the load current varies, the current taken by the diode varies by the same percentage in the opposite direction. The diode current reaches its maximum value when the load current is zero, and at this point care must be taken to ensure that the maximum power dissipation rating of the diode is not exceeded. Therefore

$$I_{D(max)} = \frac{\text{Maximum power dissipation}}{\text{Diode (output) voltage}} \tag{7.3}$$

The required value of R_1 can now be calculated using equation (7.2).

Example 7.4

A Zener diode regulator circuit is to provide a 24 V supply to a variable load. The input voltage is 30 V and a 24 V, 400 mW Zener diode is to be used. Calculate (a) the series resistance R_1 required and (b) the diode current when the load resistance is 2000 Ω.

Solution

(a) From equation (7.3),

$$I_{D(max)} = 0.4/24 = 16.67 \text{ mA}$$

From equation (7.2),

$$R_1 = (30 - 24)/(16.67 \times 10^{-3}) = 360 \ \Omega \quad (Ans.)$$

(b) When the load resistance is 2000 Ω the load current will be

$$24/2000 = 12 \text{ mA}$$

The total current in R_1 is 16.67 mA and

Diode current $= 4.67$ mA   (*Ans.*)

(b) Varying supply voltage; fixed load
If the supply voltage to the stabilizer circuit should decrease, the diode current will fall so that a smaller voltage drop across the series resistor occurs. The Zener diode must pass a minimum current $I_{D(min)}$ if it is to operate in its breakdown region and act as a voltage regulator. Therefore

$$R_1 = (V_{IN(min)} - V_{OUT})/(I_{D(min)} + I_L) \qquad (7.4)$$

Example 7.5

A 9.1 V, 1.3 W Zener diode has a minimum current requirement of 20 mA and is to be used in a voltage regulator circuit. The supply voltage is 20 V $\pm$ 10% and the constant load current is 30 mA. Calculate (*a*) the series resistance required and (*b*) the power dissipated in the diode when the supply voltage is 22 V.

Solution
    (*a*) From equation (7.4)

$$R_1 = (18 - 9.1)/[(20 + 30) \times 10^{-3}] = 178 \ \Omega \quad (Ans.)$$

(*b*) When $V_{IN} = 22$ V

$$I_D + I_L = (22 - 9.1)/178 = 72.47 \text{ mA}$$

$$I_D = 72.47 - 30 = 42.47 \text{ mA} \quad (Ans.)$$

and therefore

Power dissipated $= 9.1 \times 42.47 \times 10^{-3} = 386$ mW   (*Ans.*)

(c) Varying supply voltage; Varying load
The maximum value of the series resistor R_1 is determined by the necessary minimum diode current, and its minimum value is calculated to be large enough to ensure that the rated power dissipation of the diode is not exceeded. The value chosen for R_1 must be a compromise between the minimum and maximum values.

The voltage stability of the basic Zener diode circuit is not good enough for many applications because the internal resistance of the diode is not zero. This means that any change in the current flowing in the diode will cause a small change in the voltage appearing across the diode and hence also in the output voltage. The stability of the output voltage can be improved by using two Zener diodes as shown by Fig. 7.15.

The stabilization efficiency of the voltage regulator would be increased if the magnitude of the current flowing in the diode were reduced. The circuit of an emitter-follower voltage stabilizer is shown in Fig. 7.16. Since the transistor is connected as an emitter follower, the voltage at its emitter, which is the load voltage, is very nearly

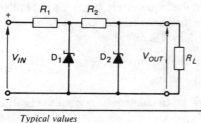

Typical values
$R_1 = 300 \ \Omega$, $R_2 = 220 \ \Omega$

Fig. 7.15 Double Zener diode stabilizer

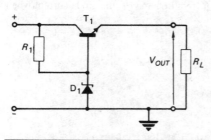

Typical value
$R_1 = 1.2 \text{ k}\Omega$

Fig. 7.16 Emitter-follower stabilizer

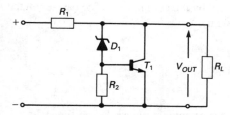

Fig. 7.17 Zener diode-parallel transistor regulator

equal to the base voltage. The base voltage is specified by the Zener diode and so the output voltage is held constant within limits determined by the diode characteristics. The emitter-follower stabilizer has no provision for varying the output voltage and its stabilization efficiency is not good enough for many applications.

An alternative circuit is shown in Fig. 7.17. The circuit operates in a similar way to the Zener diode voltage regulator but it may be used for applications where the load current is much larger than the Zener diode current. The load voltage V_{DC} is equal to $V_Z + 0.7$ volts.

Series-control Voltage Regulators

The principle of a much more efficient type of voltage regulator is shown by Fig. 7.18.

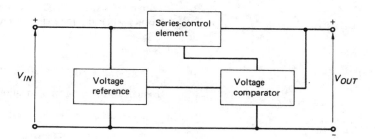

Fig. 7.18 Block diagram of series voltage regulator

The output voltage of the rectifier is applied to a series-control element which introduces resistance into the positive supply line. The output voltage V_{OUT} is smaller than the input voltage by the voltage dropped across the series element. The output voltage, or a known fraction of it, is compared in the voltage comparator with a voltage reference. The difference between the two voltages is detected and an amplified version of it is applied to the series-control element in order to vary its resistance in such a way as to maintain the output voltage at its correct value. If, for example, the output voltage is larger than it should be, the amplified difference voltage will be of such a polarity that the resistance of the controlled element will be increased and the output voltage will fall. Conversely, if the output voltage is less than its correct value, the resistance of the series-controlled element will be reduced by the amount necessary for the output voltage to rise to its correct value. Generally, the series-control element is a transistor connected as shown in Fig. 7.19. When an n-p-n transistor is employed, its collector is connected to the input terminal, and its emitter is connected to the output terminal, of the circuit since the former is more positive. If the output voltage of the stabilizer should vary by an amount δV_{OUT}, the control voltage appearing at the base terminal of the series transistor will be $A_v'\delta V_{OUT}$, which is the

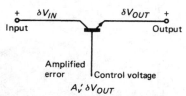

Fig. 7.19 Transistor series-control element

amplified error voltage produced by the comparator. The base-emitter voltage of the transistor is then

$$A_v' \delta V_{OUT} - \delta V_{OUT}$$

and this will produce a voltage

$$A_v'' (A_v' \delta V_{OUT} - \delta V_{OUT})$$

between the base and collector terminals where A_v'' is the voltage gain of the series transistor. The voltage across this transistor is also equal to $\delta V_{IN} - \delta V_{OUT}$ and therefore

$$\delta V_{IN} - \delta V_{OUT} = A_v'' (A_v' \delta V_{OUT} - \delta V_{OUT}) \tag{7.5}$$

Example 7.6

In a voltage stabilizer of the type shown in Fig. 7.18 the error voltage gain of the comparator is –100 and the voltage gain of the series transistor is –10. The rectifier circuit connected to the input terminals of the stabilizer has an output resistance of 200 Ω. Calculate the change in the output voltage that occurs when the load current changes by 20 mA.

Solution

$$\delta V_{IN} = \delta I_L R_{OUT} = 20 \times 10^{-3} \times 200 = 4 \text{ V}$$

Hence, substituting into equation (7.5),

$$4 - \delta V_{OUT} = -10(-100 \delta V_{OUT} - \delta V_{OUT})$$
$$4 = 1000 \delta V_{OUT} + 10 \delta V_{OUT} + \delta V_{OUT}$$
$$\delta V_{OUT} = \frac{4}{1011} = 3.956 \text{ mV} \quad (Ans.)$$

Fig. 7.20 shows the circuit of a voltage regulator in which T_1 is the series control element, T_2 is the voltage comparator and the voltage reference is provided by the Zener diode D_1. The emitter potential of T_2 is maintained at a very nearly constant value by the Zener diode D_1, while its base is held at a fraction of the output

Fig. 7.20 Transistor series voltage regulator

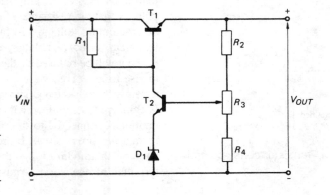

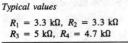

Typical values
$R_1 = 3.3 \text{ kΩ}, R_2 = 3.3 \text{ kΩ}$
$R_3 = 5 \text{ kΩ}, R_4 = 4.7 \text{ kΩ}$

voltage by the potential divider R_2, R_3 and R_4. The difference between the base and emitter potentials is amplified and the amplified error voltage is applied to the base of T_1 to vary the bias voltage provided by resistor R_1. Suppose that the output voltage of the stabilizer should increase above its nominal value (set by variable resistor R_3). The base voltage of T_2 will become more positive with respect to the constant emitter voltage, and T_2 will conduct a larger collector current. The voltage dropped across R_1 will then increase and this will make the base potential of T_1 less positive. T_1 will now conduct less readily and so its resistance increases. The consequent increase in the collector-emitter voltage of T_1 causes the output voltage to fall by an amount that is very nearly equal to the original increase. The series transistor T_1 must be capable of carrying the full load current of the stabilizer and should have an adequate power rating.

Example 7.7

The input voltage to the voltage regulator shown in Fig. 7.20 varies by 3 V. If the voltage gains of the transistors are (i) $T_1 = 20$ and (ii) $T_2 = 120$, determine the change in the output voltage. Suppose that the variable resistor R_3 is set to its middle resistance value.

Solution
Change in output voltage $= \delta V_{OUT}$
Voltage change at the base of $T_2 = \delta V_{OUT} (4.7 + 2.5)/(3.3 + 5 + 4.7)$
$$= 0.554 \, \delta V_{OUT}$$
Collector voltage of $T_1 = 120 \times 0.554 \, \delta V_{OUT}$
Base-emitter voltage of $T_1 = 120 \times 0.554 \, \delta V_{OUT} - \delta V_{OUT} = 65.48 \, \delta V_{OUT}$
Hence the changing voltage across T_1 is $20 \times 65.48 \, \delta V_{OUT} \simeq 1310 \, \delta V_{OUT}$
$$= 3 - \delta V_{OUT}$$
Therefore, $\delta V_{OUT} = 3/1311 = 2.29$ mV (*Ans.*)

Integrated Circuit Voltage Regulators

Modern electronic circuitry generally employs an IC voltage regulator to provide the required power supply voltage stability. Devices are available of varying degrees of complexity that are capable of satisfying all but the most stringent of specifications. The simplest voltage regulators, which are perfectly adequate for many applications, are three-terminal types. Representative of 3-terminal regulators are devices in the 7800 series.

The 7800 series of positive voltage regulators are able to provide an output current of up to 1.5 A with one of a number of fixed output voltages. If an excess current, or overheating, should occur the IC will shut down to prevent any damage being caused. The output voltage of a 7800 regulator is indicated by the last two figures in the device number. Thus, the 7805 provides an output voltage of 5 V, the 7808 provides 8 V, the 7812 provides 12 V and so on. The manufacturer's data sheet for the 7805 includes:

Data Sheet for the 7805

d.c. input voltage for	V_{OUT} = 5 to 18 V	35 V max.
for	V_{OUT} = 24 V	40 V max.
Quiescent current	typ. 4.2 mA,	max. 6 mA
Output voltage min.	4.8 V, typ. 5 V, max. 5.2 V	
Temperature coefficient at	I_{OUT} = 5 mA, –1.1 mV/°C	

Line regulation $\dfrac{\delta V_{IN}}{\delta V_{OUT}}$ (V_{IN} = 8 V → 12 V) 1 mV typ. 25 mV max.

Load regulation $\dfrac{\delta V_{OUT}}{V_{OUT}}$ 15 mV typ. 50 mV max.

Output resistance   17 mΩ

The IC can be used to produce a fixed output voltage in the way shown by Fig. 7.21. Capacitor C_2 is usually fitted to keep the output resistance of the circuit low at high frequencies. If the regulator is not sited immediately after the rectifier circuit itself an input capacitor C_1 is also necessary.

The 7900 series regulators provide negative voltage regulators with the same standard voltages as the 7800 series.

If there is a requirement to be able to adjust the output voltage a four-terminal voltage regulator must be employed. The fourth pin acts as a control terminal that allows the output voltage to be set to a wanted value. An example of this kind of regulator IC is shown by Fig. 7.22.

Another commonly used circuit is the 723 voltage regulator; the pin connections of the device are given by Fig. 7.23(a). An internal voltage reference provides 7.15 V at pin 6. This voltage must be connected either directly or via a potential divider to pin 5. A direct connection is used if the required output voltage is greater than 7.15 V,

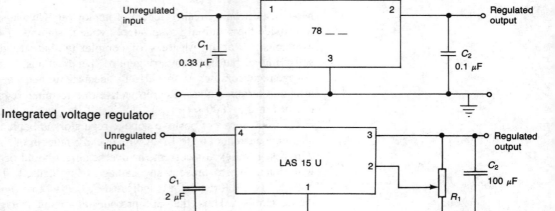

Fig. 7.21   Integrated voltage regulator

Fig. 7.22   Adjustable IC voltage regulator

adjustment of the output voltage is then achieved by means of the potential divider $R_1 + R_2$ connected across the output terminals of the circuit (see Fig. 7.23(b)). Output voltage $V_{OUT} = 7.15(R_1 + R_2)/R_2$. If the wanted output voltage is less than 7.15 V the arrangement given in Fig. 7.23(c) is employed, it can be seen that the potential divider has been moved from the output terminals to the connection between pins 5 and 6. Capacitor C_1 must be connected between the terminals 4 and 13 to prevent any possibility of high-frequency oscillations occurring. Resistor R_3 is chosen to give a voltage drop of 0.5 V when the maximum wanted output current flows; then any excess current will cause the regulator to shut down.

Dual power supplies are commonly employed to supply power to many circuits, such as op-amps, that require typically $\pm$ 12 V d.c. supply. A dual power supply such as the one shown in Fig. 7.24, consists of two separate supplies that use the same centre-tapped transformer and bridge rectifier. Each output has its own linear voltage regulator.

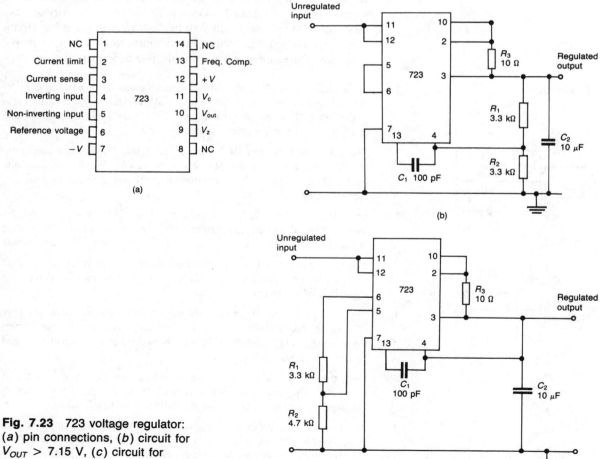

Fig. 7.23 723 voltage regulator: (a) pin connections, (b) circuit for $V_{OUT} > 7.15$ V, (c) circuit for $V_{OUT} < 7.15$ V

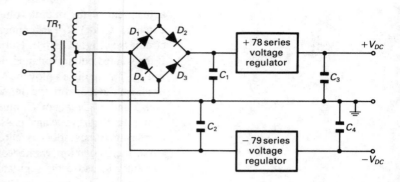

Fig. 7.24 Dual power supply

Over-voltage Protection

Over-voltage protection on the output is a feature of many power supplies. It is particularly important for the 5 V power supplies used for TTL digital circuitry. TTL devices have an absolute maximum supply voltage of 7 V and if the supply voltage were to rise above this voltage the circuit might well be destroyed. A *crow-bar* circuit is employed which effectively places a short-circuit across the output terminals if the output voltage should rise to a preset figure.

Exercises

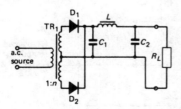

Fig. 7.25

Fig. 7.26

7.1 In the circuit of Fig. 7.25 the voltage across the secondary winding is 22 V. What should be the peak inverse voltage of each diode? What is the value of the voltage V_S when the current in the 150 Ω load is 50 mA? Assume that 0.5 V is dropped across the conducting diode.

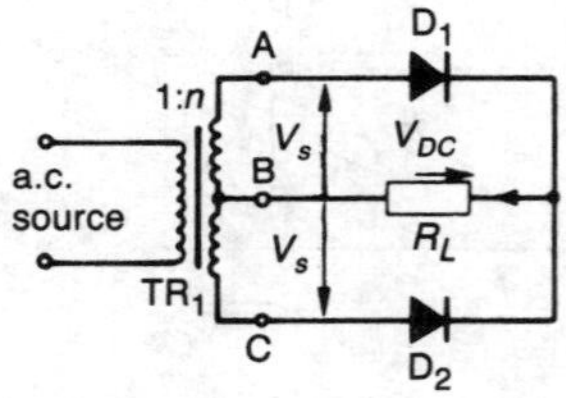

7.2 The input transformer of a full-wave rectifier (Fig. 7.25) has a turns ratio of 9.58:1. The r.m.s. voltage at the secondary is 24 V. Calculate (a) the r.m.s. input voltage, (b) the peak current flowing in the 200 Ω load and (c) the d.c. current in the 200 Ω load.

7.3 Draw the circuit of a half-wave rectifier circuit. If the voltage across the secondary winding of the transformer is 60 V what is the peak inverse voltage of the diode when the reservoir capacitor is (a) connected, (b) disconnected?

7.4 The no-load output voltage of a rectifier circuit is 24 V. When the full-load current is taken the output voltage falls to 23.4 V. Calculate the percentage regulation of the circuit.

7.5 Fig. 7.26 shows the circuit of a full-wave rectifier circuit. Draw sketches to show the expected waveforms (a) at the primary winding of the transformer, (b) at the secondary winding of the transformer, (c) at the junction of C_1 and L_1 and (d) across the load.

7.6 Draw the circuit of a half-wave rectifier with a capacitor input LC filter. If such a circuit has a no-load output voltage of 100 V and a regulation of 0.94% calculate the output voltage when the full load current is taken.

7.7 The circuit shown in Fig. 7.27 is to be used to produce 12 V output voltage at a current of 120 mA. The input voltage may vary between 18 V and 20 V. Calculate the necessary value of the series resistor R_s if the diode minimum current is 5 mA.

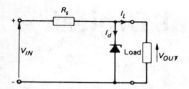

Fig. 7.27

7.8 In the circuit of Fig. 7.27 two Zener diodes are available. One of the diodes has a maximum power dissipation of 0.6 W and the other dissipates 1.2 W. Can either diode be used in the circuit?

7.9 A 24 V stabilized voltage is to be obtained from a 30 V d.c. supply. A 24 V, 3 W Zener diode is to be used. Calculate the required value for the series resistor.

7.10 A Zener diode stabilizing circuit has an input voltage of 18 V and a diode current of 8 mA to give 10 V across the load of 1200 ohms. Calculate the value of the series resistor and the diode current when the load resistance is 1000 ohms.

7.11 A 14.2 V Zener diode has a maximum power dissipation of 1.5 W. Calculate the maximum current that the diode may conduct.

8 Combinational Digital Circuits

Modern electronics make ever increasing use of digital electronic circuitry which responds only to signals that can only take up either one of two logic levels. Either the signal is HIGH or it is LOW. In most circuits the HIGH state is used to represent binary 1 while the LOW state represents binary 0. For example, in the TTL logic family binary 0 is indicated by a voltage in the range 0 V to + 0.8 V and binary 1 is indicated by a voltage in the range + 2 V to + 5 V. For CMOS devices logic 0 is less than $V_{DD}/3$ and logic 1 is greater than $2V_{DD}/3$, where V_{DD} is the power supply voltage.

The Binary Code

In digital electronic systems, the devices have two stable states, ON and OFF, and for this reason the binary number system is used. In the binary system only two digits 0 and 1 exist. Larger numbers are obtained by utilizing the powers of two. The digit at the right-hand side of a binary number represents a multiple (0 or 1) of 2^0; the next digit to the left represents a multiple of 2^1; and so on as shown by Table 8.1. The binary digits are generally known as bits.

The value of each power of two is given in the table and any desired number can be attained by a correct choice of zeros and ones. Thus, number 18 for example: 18 is equal to 16 plus 2 and is therefore given by 0010010 in a 7-bit code or by 10010 if only 5 bits are used. Reading from the right (the *least significant bit* or *lsb*), the number consists of zero 1, one 2, zero 4, zero 8 and one 16. Similarly, the binary equivalents of some other numbers are given in Table 8.2, assuming an 8-bit code.

Some typical examples of digital systems are the following: (*a*) the digital computer, and communication with a computer over telephone

Table 8.1 Binary code

2^8	2^7	2^6	2^5	2^4	2^3	2^2	2^1	2^0
256	128	64	32	16	8	4	2	1

Table 8.2 Binary equivalents of decimal numbers

7 00000111	25 00011001	80 01010000
17 00010001	31 00011111	150 10010110

lines; (*b*) the control of traffic lights, of lifts in tall buildings, of conveyor belts in factories, etc.; (*c*) safety arrangements for cranes and various factory machines; (*d*) electronic counting systems such as counting the number of cars in a car park; (*e*) digital voltmeters; (*f*) digital watches; and (*g*) pocket calculators.

Combinational Logic

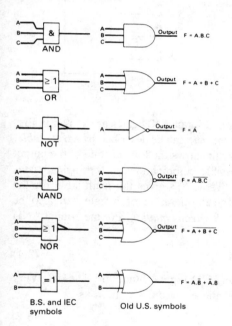

B.S. and IEC symbols

Old U.S. symbols

Fig. 8.1 Gate symbols
(▷ indicates an active-low output)

Combinational logic circuits are digital circuits the states of whose outputs depend on the present logic state of the inputs. A combinational logic circuit is unable to store any information, a combinational logic circuit is often made by the interconnection of a number of *gates*

An electronic gate is a logic element which is able to operate on an applied binary digital signal in a manner determined by its logical function. Seven different types of gate feature in digital circuitry, these are (i) the AND gate, (ii) the OR gate, (iii) the NOT gate, (iv) the NOR gate, (v) the NAND gate, (vi) the exclusive-OR gate and (vii) the exclusive-NOR gate.

The British Standards Institution (BSI) and International Electrotechnical Commission (IEC) symbols for the various gates are given in Fig. 8.1. The old American gate symbols which are often used are also shown in Fig. 8.1.

Modern logic circuitry is invariably produced using integrated circuit technology. The majority of digital ICs belong to either the *CMOS* or the *TTL* logic families. The latter family has several subgroups of which the most popular is known as low-power Schottky TTL. The use of ICs for electronic equipment results in greater reliability and lower costs and allows complex circuit functions to be economically produced. Because of this, complex circuitry, such as is used in quartz watches, pocket calculators and home computers, has become relatively inexpensive and is commonly employed.

In both the CMOS and the TTL families, a standard package may accommodate more than one gate.

The NOT Gate

The NOT gate, or inverter, is used to invert a logic term, e.g. change logical 0 into logical 1 or to change 1 to 0. In a Boolean expression the NOT logical function is indicated by a bar placed over a symbol. Thus $\bar{A}$ means 'NOT A'.

The NOT function is provided in both the CMOS and the TTL logic families by a hex inverter. This device, for example the TTL 7404, consists of six independent inverters or NOT gates in the one IC package.

The AND Gate

The AND gate is a logic circuit having two, or more, input terminals, labelled A, B, C, etc., and a single output terminal, usually labelled F. The output F of an AND gate is HIGH, or at logical 1, only if *all* of its inputs are HIGH, or at logical 1. If any one, or more, of the inputs are LOW, or at logical 0, then the output of the circuit will be LOW, or at logical 0.

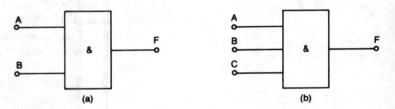

Fig. 8.2 (*a*) 2-input and (*b*) 3-input AND gates

The symbols for 2-input and 3-input AND gates are shown in Fig. 8.2. The logical operation of the gate can be described by a *truth table*. The truth table of a logic circuit shows the logical state of the output of the circuit for all the possible combinations of the logical states of the inputs to the circuit. Table 8.3 gives the truth table for the 2-input AND gate. The logical operation of a gate can also be described by writing down its *Boolean equation*. For the 2-input AND gate this equation is

$$F = A.B \qquad (8.1)$$

The Boolean symbol for the AND logical function is the dot. Very often the dot is omitted and the equation is written as

$$F = AB \qquad (8.2)$$

The number of columns in a truth table is equal to 2^n, where n is the number of inputs. For the 2-input gate $n = 2$ and so there are 2^2 or 4 columns in Table 8.3. The truth table for a 3-input AND gate

Table 8.3 2-input AND gate

A	0	1	0	1
B	0	0	1	1
F	0	0	0	1

Table 8.4 3-input AND gate

A	0	1	0	1	0	1	0	1
B	0	0	1	1	0	0	1	1
C	0	0	0	0	1	1	1	1
F	0	0	0	0	0	0	0	1

must have 2^3 or 8 columns and it is given by Table 8.4. The Boolean equation for the 3-input AND gate is given by equation (8.3), i.e.

$$F = ABC \qquad (8.3)$$

Several different TTL circuits provide the AND gate function:

7408: quad 2-input AND gates
7411: triple 3-input AND gates
7421: dual 4-input AND gates.

The pin connections of the first two of these ICs are given in Fig. 8.11.

Example 8.1

(*a*) How many input combinations are there for a 4-input AND gate? (*b*) How many inputs must be (i) at binary 1 for the output to also be at 1, (ii) at binary 0 for the output to be at 0? (*c*) Write down the Boolean equation for the gate.

Solution
(*a*) Number of input combinations = 2^4 = 16   (*Ans.*)
(*b*) (i) All four. (ii) Just one   (*Ans.*)
(*c*) $F = ABCD$ (*Ans.*)

The OR Gate

The OR gate, Fig. 8.3, has its output F at the logical 1 level whenever any one, or more, of its inputs are at logical 1. The output of the gate will be at logical 0 only if all of its inputs are at logical 0. Table 8.5 gives the truth tables for 2-input and 3-input OR gates. The Boolean equations for each gate are given by equations (8.4) and (8.5) respectively, i.e.

$$F = A + B \qquad (8.4)$$

$$F = A + B + C \qquad (8.5)$$

Note that the symbol for the OR logical function is the + sign.

There is only one OR gate in most of the TTL logic sub-families, and this is the 7432 quad 2-input OR gate. If a 3-input or a 4-input OR gate is wanted then two 2-input OR gates must be combined as

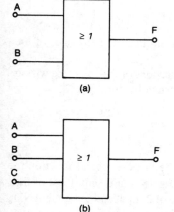

Fig. 8.3 2-input, and 3-input OR gates

Table 8.5 2-input and 3-input OR gates

A	0	1	0	1		A	0	1	0	1	0	1	0	1
B	0	0	1	1		B	0	0	1	1	0	0	1	1
F	0	1	1	1		C	0	0	0	0	1	1	1	1
						F	0	1	1	1	1	1	1	1

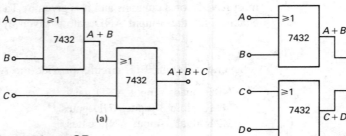

Fig. 8.4 Use of (a) two 2-input OR gates to form a 3-input OR gate (b) three two-input OR gates to form a 4-input gate

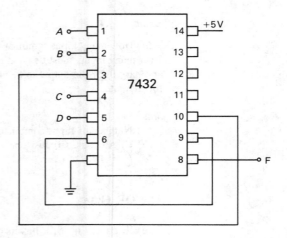

Fig. 8.5 4-input OR gate implemented using the TTL 7432 dual 2-input OR gate

shown by Fig. 8.4. In figure (a) two, and in figure (b) one, of the OR gates is left unused.

Referring to Fig. 8.11 for the pin connections of the 7432, the wiring of the 4-input OR gate circuit is as shown by Fig. 8.5.

The NAND Gate

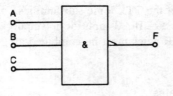

Fig. 8.6 NAND gate

The NAND gate, Fig. 8.6, performs the inverse logical function to the AND gate. The output of a NAND gate is at 0 only if all of the inputs to the gate are at 1. The truth table of 2-input and 3-input NAND gates are given by Table 8.6. The Boolean expressions for the two NAND gates are given by equations (8.6) and (8.7) respectively.

$$F = \overline{AB} \qquad (8.6)$$

$$F = \overline{ABC} \qquad (8.7)$$

The NOT function can be produced using a NAND gate in either of two different ways. Referring to the truth table of the 2-input NAND gate it can be seen that:

Table 8.6 2-input and 3-input NAND gates

A	0	1	0	1	A	0	1	0	1	0	1	0	1
B	0	0	1	1	B	0	0	1	1	0	0	1	1
F	1	1	1	0	C	0	0	0	0	1	1	1	1
					F	1	1	1	1	1	1	1	0

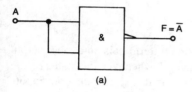

(a)

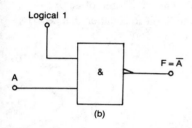

(b)

Fig. 8.7 NAND gate connected as an inverter

(a) if the inputs A and B are connected together so that $A = B$, Fig. 8.7(a), the output will always be inverted, and

(b) if either input is held at the logical 1 voltage level, Fig. 8.7(b), the output will always be NOT the logical state of the other input.

There are a number of NAND gates available in the various TTL sub-families:

7400 quad 2-input NAND gates
7410 triple 3-input NAND gates
7420 dual 4-input NAND gates
7430 single 8-input NAND gate.

The NOR Gate

The NOR gate, Fig. 8.8, performs the inverse of the logical OR function. This means that the output F of a NOR gate is at logical 1 only when all of its inputs are at logical 0. The truth tables for both 2-input and 3-input NOR gates are given by Table 8.7. The Boolean expressions for the two gates are given by

$$F = \overline{A + B} \tag{8.8}$$

$$F = \overline{A + B + C} \tag{8.9}$$

The NOT function can be generated by a NOR gate if either (a) both its input terminals are connected together, Fig. 8.9(a), or (b) one input is connected to the logical 0 voltage level, Fig. 8.9(b). The NOR gates in the TTL family are:

7402 quad 2-input NOR gates
7427 triple 3-input NOR gates.

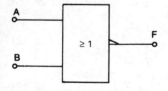

Fig. 8.8 NOR gate

Table 8.7 2-input and 3-input NOR gates

A	0	1	0	1	A	0	1	0	1	0	1	0	1
B	0	0	1	1	B	0	0	1	1	0	0	1	1
F	1	0	0	0	C	0	0	0	0	1	1	1	1
					F	1	0	0	0	0	0	0	0

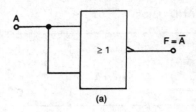

(a)

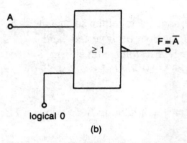

logical 0

(b)

Fig. 8.9 NOR gate connected as an inverter

Table 8.8 Exclusive OR

A	0	1	0	1
B	0	0	1	1
F	0	1	1	0

Table 8.9 Exclusive NOR

A	0	1	0	1
B	0	0	1	1
F	1	0	0	1

Digital Integrated Circuits

Example 8.2

The inputs to a 3-input NOR gate are two at binary 0 and the third at binary 1. What is the logical state of the output of the gate?

Solution

$F = \overline{A + B + C}$ so if any of the inputs is at 1 the output will be at 0. Therefore, $F = 0$ (*Ans.*)

Exclusive-OR Gate

The exclusive-OR gate has two input terminals and one output terminal. The output will be at binary 1 whenever the states of the two inputs are not the same. If both inputs are at 1 or both inputs are at 0 the output of the gate is at 0. The truth table of an exclusive-OR gate is given by Table 8.8, and the Boolean equation that describes the operation of the gate is given by

$$F = A\bar{B} + \bar{A}B \qquad (8.10)$$

This is often written as

$$F = A \oplus B \qquad (8.11)$$

Exclusive-NOR Gate

An exclusive-NOR gate has two input terminals and one output terminal. The output will be at binary 1 only when both inputs are at the same logic state. Whenever the inputs are at different logical states the output will be at logic 0. The truth table of an exclusive-NOR gate is given by Table 8.9 and the Boolean expression for the gate is given by equation (8.12).

$$F = AB + \bar{A}\bar{B} \qquad (8.12)$$

This is often written as

$$F = \overline{A \oplus B} \qquad (8.13)$$

Example 8.3

The waveforms given in Fig. 8.10(*a*) are applied to the input terminals of a 2-input gate. Draw the output waveform if the gate is (*a*) AND; (*b*) OR; (*c*) NAND; (*d*) NOR; (*e*) exclusive-OR or (*f*) exclusive-NOR.

Solution

The output waveforms are shown by Fig. 8.10(*b*).

In practice, the vast majority of the gates employed in digital circuitry are members of either the TTL or the CMOS logic families. Some TTL examples, showing IC package pin connections are given in Fig.

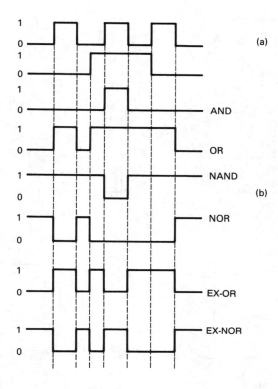

Fig. 8.10

8.11(*a*). The TTL family includes the standard 74 series, the low-power Schottky 74LS series, the advanced Schottky 74AS series and the advanced low-power Schottky 74ALS series. The examples given in Fig. 8.11(*a*) have representatives in each series.

The CMOS logic family includes the standard 4000 series, the high-speed HCMOS series and the advanced CMOS series. Some examples of the 4000 series gates are given by Fig. 8.11(*b*).

It is normally preferable for economic reasons for one circuit to employ ICs from only one logic family. The choice of the logic family is based on a number of factors, such as power dissipation and speed of operation.

Combinations of Gates

Very often it is convenient to be able to perform a logical function using a different kind of gate.

AND Followed by NOT

If an AND gate is followed by a NOT gate (see Fig. 8.12) the output of the circuit will be $\overline{AB}$. Therefore, the combination acts like a NAND gate.

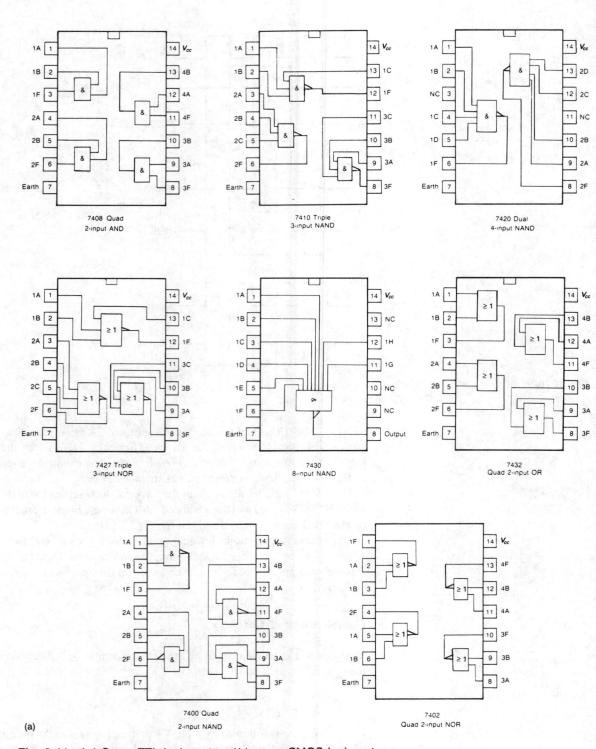

7408 Quad
2-input AND

7410 Triple
3-input NAND

7420 Dual
4-input NAND

7427 Triple
3-input NOR

7430
8-input NAND

7432
Quad 2-input OR

7400 Quad
2-input NAND

7402
Quad 2-input NOR

(a)

Fig. 8.11 (a) Some TTL logic gates, (b) some CMOS logic gates

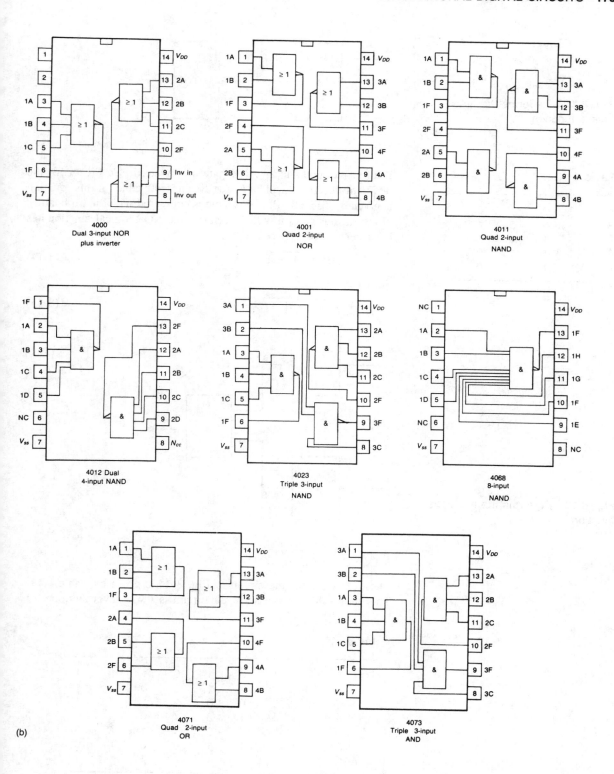

4000
Dual 3-input NOR
plus inverter

4001
Quad 2-input
NOR

4011
Quad 2-input
NAND

4012 Dual
4-input NAND

4023
Triple 3-input
NAND

4068
8-input
NAND

4071
Quad 2-input
OR

4073
Triple 3-input
AND

(b)

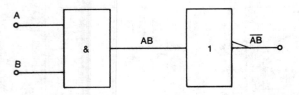

Fig. 8.12 Performs the NAND logical function

Table 8.10

A	0	1	0	1
B	0	0	1	1
$\bar{A}$	1	0	1	0
$\bar{B}$	1	1	0	0
$F = \bar{A}\bar{B}$	1	0	0	0

Fig. 8.13 Performs the NOR logical function

AND Preceded by NOT

If all of the inputs to an AND gate are inverted, see Fig. 8.13, the output of the circuit will be $F = \bar{A}.\bar{B}$. The truth table for this equation is given by Table 8.10. If this table is compared with the truth tables of the various gates it will be seen that the NOR logical function has been performed. Therefore,

$$F = \bar{A}\bar{B} = \overline{A + B} \tag{8.14}$$

Similarly,

$$F = \bar{A}\bar{B}\bar{C} = \overline{A + B + C} \tag{8.15}$$

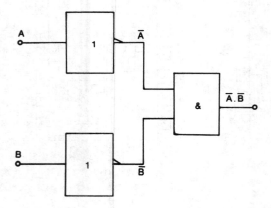

NAND Preceded by NOT

The truth table for the circuit of Fig. 8.14 is given by Table 8.11. This shows that the output of the circuit is 1 whenever any one, or

Table 8.11

A	0	1	0	1	0	1	0	1
B	0	0	1	1	0	0	1	1
C	0	0	0	0	1	1	1	1
$\bar{A}$	1	0	1	0	1	0	1	0
$\bar{B}$	1	1	0	0	1	1	0	0
$\bar{C}$	1	1	1	1	0	0	0	0
$F = \overline{\bar{A}\bar{B}\bar{C}}$	0	1	1	1	1	1	1	1

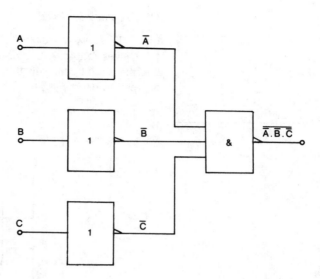

Fig. 8.14 Performs the OR logical function

more, of its inputs is at 1. This means that the circuit performs the OR logical function. The Boolean expression for the circuit is

$$F = \overline{\overline{A}\overline{B}\overline{C}} = A + B + C \tag{8.16}$$

Similarly,

$$F = \overline{ABC} = \bar{A} + \bar{B} + \bar{C} \tag{8.17}$$

NAND Followed by NOT

When a NAND gate is followed by a NOT gate, Fig. 8.15, the AND function is performed.

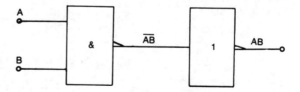

Fig. 8.15 Performs the AND logical function

OR Followed by NOT

Fig. 8.16 shows a 2-input OR gate that is followed by a NOT gate. The combination acts like a NOR gate.

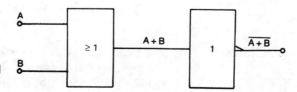

Fig. 8.16 Performs the NOR logical function

Table 8.12

A	0	1	0	1
B	0	0	1	1
$\bar{A}$	1	0	1	0
$\bar{B}$	1	1	0	0
$F = \bar{A} + \bar{B}$	1	1	1	0

OR Preceded by NOT

Fig. 8.17 shows a 2-input OR gate that has both of its inputs fed via NOT gates. Table 8.12 gives the truth table of the circuit and shows that the circuit performs the logical function NAND. The Boolean expression describing the circuit is given by equation (8.18), i.e.

$$F = \bar{A} + \bar{B} = \overline{AB} \qquad (8.18)$$

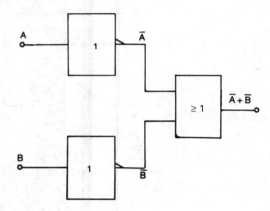

Fig. 8.17 Performs the NAND logical function

NOR Followed by NOT

Fig. 8.18 shows a NOR gate followed by an inverter. The OR logical function is performed.

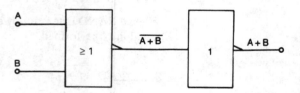

Fig. 8.18 Performs the OR logical function

NOR Preceded by NOT

Fig. 8.19 shows a NOR gate both of whose inputs are passed through a NOT gate. The truth table of the arrangement is given by Table 8.13. The output of the circuit is at logical 1 only when both of its inputs are at logical 1. Hence, the circuit performs the AND logical function. The Boolean expression for the circuit is given by equation (8.19), i.e.

Table 8.13

A	0	1	0	1
B	0	0	1	1
$\bar{A}$	1	0	1	0
$\bar{B}$	1	1	0	0
$F = \overline{\bar{A} + \bar{B}}$	0	0	0	1

$$F = \overline{\bar{A} + \bar{B}} = AB \qquad (8.19)$$

Combinational Logic Circuits

The basic logic gates are used in combination to form other, more complex, logical functions. These logic functions are formed by suitably inter-connecting the individual gates. Some of the functions

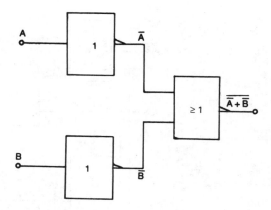

Fig. 8.19 Performs the AND logical function

that may be required can also be obtained 'ready-made', as it were, in an SSI package. Logic functions can also be generated using LSI circuits known as ROMs and PLDs but only the use of the basic gates will be considered here.

Analysis

The function of an existing combination logic circuit is expressed by the Boolean equation that describes the signal that appears at its output terminal(s). The output function can be determined by following through the circuit the output of each gate as it is connected to the inputs of one or more other gates. An example of this is given by the circuit shown in Fig. 8.20. The output of the upper AND gate is AB and the output of the lower AND gate is BC. These two outputs are applied to the input of the OR gate. Hence, the output of the circuit is $F = AB + BC$. The truth table of the circuit is given by Table 8.14.

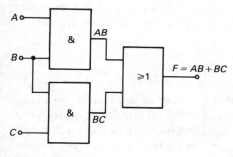

Fig. 8.20

Table 8.14

A	0	1	0	1	0	1	0	1
B	0	0	1	1	0	0	1	1
C	0	0	0	0	1	1	1	1
AB	0	0	0	1	0	0	0	1
BC	0	0	0	0	0	0	1	1
F	0	0	0	1	0	0	1	1

A rather more complex circuit is shown in Fig. 8.12. The upper NAND gate has $\bar{A}$ and B as its inputs and so its output is $\overline{\bar{A}B}$. Similarly, the output of the lower NAND gate is $\overline{A\bar{B}}$. These two outputs are applied to the inputs of the third NAND gate. The output of the circuit is $\overline{\overline{\bar{A}B}.\overline{A\bar{B}}}$. This Boolean equation looks rather complicated but perhaps it can be simplified. The truth table for the circuit is given by Table 8.15.

Table 8.15

A	0	1	0	1
$\bar{A}$	1	0	1	0
B	0	0	1	1
$\bar{B}$	1	1	0	0
$A\bar{B}$	0	1	0	0
$\bar{A}B$	0	0	1	0
$\overline{A\bar{B}}$	1	0	1	1
$\overline{\bar{A}B}$	1	1	0	1
$\overline{A\bar{B}}.\overline{\bar{A}B}$	1	0	0	1
$\overline{\overline{A\bar{B}}.\overline{\bar{A}B}}$	0	1	1	0

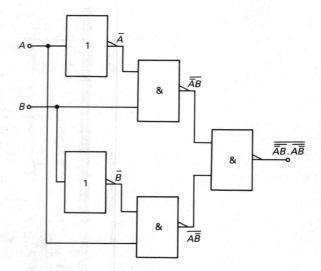

Fig. 8.21

If the last line of the table is compared with the last line of Table 8.8 it will be seen that they are identical. This means that the circuit given in Fig. 8.21 performs the logical function exclusive-OR.

Design

Many digital circuits consist of a number of gates which have been interconnected to perform a wanted logical function. The design of a combinational logic circuit starts with the truth table which describes the required logical operation. Once the truth table has been written down it can be used to derive the Boolean expression which describes the circuit. The expression should contain one term for each column (or row) in the truth table for which the output F of the circuit is at the logical 1 state. The expression so obtained should then be simplified, or *minimized*, if possible to eliminate all unnecessary terms. This step ensures that the final circuit contains the minimum number of gates. Next, the designed circuit is usually converted into an equivalent circuit that uses *either* NAND *or* NOR gates only. Besides very often reducing the number of integrated circuits required, the use of just one kind of gate makes both manufacture and fault-finding somewhat easier.

Except for the most simple problems the Boolean equation will not be in its most simplified form. Usually, it will contain a number of redundant terms and so it is often necessary to first simplify each derived Boolean equation. There are two main ways in which Boolean equations can be reduced to their most simple form: one is an algebraic method which relies on a number of Boolean laws and identities; the other uses a mapping technique known as a Karnaugh map.

Boolean Laws and Identities

The more commonly used Boolean laws and identities are listed in Table 8.16. Each of these laws and identities can be proved by writing down the various combinations of A, B and C in a truth table.

Table 8.16

1.	$A + B = B + A$
2.	$AB = BA$
3.	$(A + B) + C = A + (B + C) = (A + C) + B = A + B + C$
4.	$(AB)C = ABC$
5.	$A(B + C) = AB + AC$
6.	$AA = A$
7.	$\bar{\bar{A}} = A$
8.	$A + A = A$
9.	$A + \bar{A} = 1$
10.	$A + 0 = A$
11.	$A + 1 = 1$
12.	$A1 = A$
13.	$A0 = 0$
14.	$A + AB = A(1 + B) = A$
15.	$(A + B)(A + C) = AA + AC + AB + BC = A(1 + B + C) + BC = A + BC$
16.	$A + \bar{A}B = A + B$

Example 8.4

Use the Boolean identities given in Table 8.16 to simplify each of the following Boolean equations.

(a) $F = ABC + \overline{ABC}$, (b) $F = \overline{\overline{ABC}} + A$, (c) $F = \bar{A}\bar{B}\bar{C} + \bar{A}\bar{B}C$,
(d) $F = AB + A\bar{B} + \bar{A}B$, (e) $F = AB\bar{C} + \bar{A}BC + ABC$, and
(f) $F = A\bar{B}C + \bar{A}BC + \bar{A}\bar{B}C + ABC$

Solution

(a) From (9), $F = 1$ (*Ans.*)
(b) $F = \overline{\overline{ABC}} + A = A + ABC = A(1 + BC)$. From (11), $F = A$ (*Ans.*)
(c) $F = \bar{A}\bar{B}(C + \bar{C})$. From (9), $F = \bar{A}\bar{B}$ (*Ans.*)
(d) $F = A(B + \bar{B}) + \bar{A}B = A + \bar{A}B = A + B$ (*Ans.*)
(e) $F = AB(C + \bar{C}) + \bar{A}BC = AB + \bar{A}BC$
   $= B(A + \bar{A}C) = B(A + C) = AB + BC$ (*Ans.*)
(f) $F = BC(A + \bar{A}) + \bar{B}C(A + \bar{A}) = BC + \bar{B}C = C(B + \bar{B}) = C$ (*Ans.*)

There are also several other identities that, although sometimes useful, are not essential to know, plus two *very* important rules that are known as *De Morgan's Rules*.

De Morgan's Rules

The two rules attributed to De Morgan are

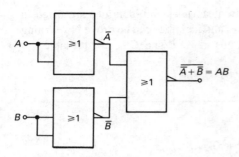

Fig. 8.22 The AND function implemented using NOR gates

A $\overline{AB} = \bar{A} + \bar{B}$ (8.20)

B $\overline{A + B} = \bar{A}\bar{B}$ (8.21)

Both of these rules have already been arrived at by the use of truth tables, see Tables 8.11 and 8.13. The rules governing the application of De Morgan's theorems are:

(a) invert the variables
(b) change the connections
(c) invert the whole expression.

Since either of De Morgan's rules produces the equivalent of any expression to which it is applied it can be applied to any part of, or the whole of, a Boolean expression. The rules still apply when there are three variables. Thus:

$$\overline{ABC} = \bar{A} + \bar{B} + \bar{C}$$
$$\overline{A + B + C} = \bar{A}\bar{B}\bar{C}$$

Example 8.5

Simplify $F = \overline{AB(B + C)}$

Solution

$F = \overline{AB} + \overline{B + C} = \bar{A} + \bar{B} + \bar{B}\bar{C} = \bar{A} + \bar{B}(1 + \bar{C})$
    $= \bar{A} + \bar{B}$ (*Ans.*)

Example 8.6

Simplify $F = \overline{(A + B)C} + \overline{AB + C}$

Solution

$F = \overline{A + B} + \bar{C} + \overline{AB}.\bar{C}$
    $= \bar{A}\bar{B} + \bar{C} + (\bar{A} + \bar{B})\bar{C}$
    $= \bar{A}\bar{B} + \bar{C}(1 + \bar{A} + \bar{B})$
    $= \bar{A}\bar{B} + \bar{C}$ (*Ans.*)

Example 8.7

Simplify $F = \overline{((\overline{A + B})C)(AB + C)}$

Solution

$F = (\overline{A + B} + \bar{C})(\overline{AB}.\bar{C})$
    $= (\bar{A}\bar{B} + \bar{C})(\bar{A} + \bar{B})\bar{C}$
    $= \bar{A}\bar{B}\bar{C} + \bar{A}\bar{B}\bar{C} + \bar{A}\bar{C} + \bar{B}\bar{C}$
    $= \bar{A}\bar{B}\bar{C} + \bar{A}\bar{C} + \bar{B}\bar{C}$
    $= \bar{A}\bar{C}(1 + \bar{B}) + \bar{B}\bar{C}$
    $= \bar{A}\bar{C} + \bar{B}\bar{C}$ (*Ans.*)

Example 8.8

Simplify $F = \overline{\bar{A} + B + C} + A\bar{B}\bar{C}$

Solution

$F = A\bar{B}\bar{C} + A\bar{B}\bar{C} = A\bar{B}\bar{C}$ (*Ans.*)

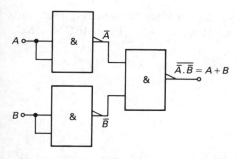

Fig. 8.23 The OR function implemented using NAND gates

The Use of NAND and NOR Gates to Generate the AND and OR Logical Functions

It is common practice to use either NAND or NOR gates only in a combinational logic circuit. The use of both types of gate to generate the NOT function and of two NAND gates to generate the AND function, or two NOR gates to give the OR function is described on page 175. The use of De Morgan's rules shows that it is also possible to produce the AND function using NOR gates only and the OR function using only NAND gates.

One rule is $\overline{AB} = \bar{A} + \bar{B}$ and hence $AB = \overline{\bar{A} + \bar{B}}$. Fig. 8.22 shows that this equation can be implemented using three NOR gates.

The other rule is $\overline{A + B} = \bar{A}\bar{B}$ and hence $A + B = \overline{\bar{A}\bar{B}}$, and this can easily be implemented using three NAND gates as shown by Fig. 8.23.

In both cases the concept can be extended to three or more input variables.

The Karnaugh Map

The Karnaugh map is a graphical representation of *all* the combinations of the input variables that can exist in a logical circuit. The map consists of a number of squares, each of which represents a unique combination of the input variables. The number of squares in a map must be equal to 2^n, where n is the number of input variables.

This means that a Boolean expression with two variables must be represented by a 4-square map, and an expression with three variables by an 8-square map, as shown.

	$\bar{A}$	A
$\bar{B}$	$\bar{A}\bar{B}$	$A\bar{B}$
B	$\bar{A}B$	AB

	$\bar{A}$		A	
$\bar{C}$	$\bar{A}\bar{B}\bar{C}$	$\bar{A}B\bar{C}$	$AB\bar{C}$	$A\bar{B}\bar{C}$
C	$\bar{A}\bar{B}C$	$\bar{A}BC$	ABC	$A\bar{B}C$
	$\bar{B}$	B		$\bar{B}$

To map an equation the equation should first be put into the sum-of-products form, e.g. $F = ABC + \bar{A}B\bar{C} + A\bar{B}\bar{C}$. Each term in the equation is mapped by a 1 in the appropriate square. Each term that is not present in the equation is mapped by a 0 in the appropriate square. Thus for the expression quoted previously the mapping is:

	$\bar{A}$		A	
$\bar{C}$	0	1	0	1
C	0	0	1	0
	$\bar{B}$	B		$\bar{B}$

A term like AB is mapped by putting a 1 into two squares. The term can be rewritten as $AB(C+\bar{C})$ (since $C+\bar{C}=1$) or $ABC + AB\bar{C}$ and so the mapping is:

	$\bar{A}$		A	
$\bar{C}$	0	0	1	0
C	0	0	1	0
	$\bar{B}$	B		$\bar{B}$

A single variable, such as B for example, is mapped by putting 1 into four squares; B can be rewritten as $B(A+\bar{A})\,(C+\bar{C}) = ABC + AB\bar{C} + \bar{A}BC + \bar{A}B\bar{C}$ and so the mapping is:

	$\bar{A}$		A	
$\bar{C}$	0	1	1	0
C	0	1	1	0
	$\bar{B}$	B		$\bar{B}$

A Karnaugh mapping of a Boolean expression can be used to minimize the expression by looping together 'adjacent' squares in groups of *two, four* or *eight*. Two squares are considered to be adjacent if they are (*a*) side-by-side, either horizontally or vertically, but *not* diagonally, (*b*) at each end of the map and in the same row. Thus, squares 0 and 2, 0 and 4, and 0 and 1 are adjacent but 1 and 7 are not. When squares are looped together all terms of the form $A+\bar{A}$ become equal to 1 and are redundant.

	$\bar{A}$		A	
$\bar{C}$	0	2	3	1
C	4	6	7	5
	$\bar{B}$	B		$\bar{B}$

Example 8.9

Use a Karnaugh map to simplify the equations (*a*) $F = ABC + A\bar{B}C + \bar{A}BC$ (*b*) $F = AB + \bar{B}C + \bar{A}C$ and (*c*) $F = A\bar{C} + \bar{A}B + \bar{A}\bar{B}C + \bar{A}\bar{B}\bar{C}$.

Solution
  (*a*) The mapping is:

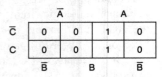

	$\bar{A}$		A	
$\bar{C}$	0	0	0	0
C	0	1	1	1
	$\bar{B}$	B		$\bar{B}$

The looped squares give $F = AC + BC$ (*Ans.*)

(b) The mapping is:

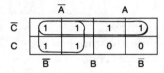

From the map, $F = AB + C$ (*Ans.*)

(c) The mapping is:

$$
\begin{array}{c|c|c|c|c|}
 & \multicolumn{2}{c}{\overline{A}} & \multicolumn{2}{c}{A} \\
\hline
\overline{C} & 1 & 1 & 1 & 1 \\
\hline
C & 1 & 1 & 0 & 0 \\
\hline
 & \overline{B} & B & \overline{B} &
\end{array}
$$

From the map, $F = \overline{A} + \overline{C}$ (*Ans.*)

Example 8.10

Use a Karnaugh map to simplify the equation $F = \overline{AB}C + \overline{A}\overline{C} + (A + C)(B + \overline{A})$. Confirm the result with a truth table.

Solution

First apply De Morgan's rules.
$$F = (\overline{A\overline{B}C})(\overline{A\overline{C}}) + \overline{A + C} + \overline{B + \overline{A}}$$
$$= (\overline{A} + B + \overline{C})(A + C) + \overline{A}C + \overline{B}A$$
$$= \overline{A}C + AB + BC + A\overline{C} + \overline{A}\overline{C} + A\overline{B}$$

The mapping is:

$$
\begin{array}{c|c|c|c|c|}
 & \multicolumn{2}{c}{\overline{A}} & \multicolumn{2}{c}{A} \\
\hline
\overline{C} & 1 & 1 & 1 & 1 \\
\hline
C & 1 & 1 & 1 & 1 \\
\hline
 & \overline{B} & B & \overline{B} &
\end{array}
$$

From the map, $F = 1$ (*Ans.*)
The truth table is shown by Table 8.17.

Table 8.17

A	B	C	$\overline{A}$	$\overline{B}$	$\overline{C}$	$\overline{A}C$	AB	BC	$A\overline{C}$	$\overline{A}\overline{C}$	$A\overline{B}$	F
0	0	0	1	1	1	0	0	0	0	1	0	1
1	0	0	0	1	1	0	0	0	1	0	1	1
0	1	0	1	0	1	0	0	0	0	1	0	1
1	1	0	0	0	1	0	1	0	1	0	0	1
0	0	1	1	1	0	1	0	0	0	0	0	1
1	0	1	0	1	0	0	0	0	0	0	1	1
0	1	1	1	0	0	1	0	1	0	0	0	1
1	1	1	0	0	0	0	1	1	0	0	0	1

Logic Circuits: Design Examples

Table 8.18

A	0	1	0	1	0	1	0	1
B	0	0	1	1	0	0	1	1
C	0	0	0	0	1	1	1	1
F	0	1	0	1	1	1	0	1

Table 8.18 gives the truth table of a combinational logic circuit that is to be designed.

The Boolean expression describing the logical operation of the circuit is $F = A\bar{B}\bar{C} + AB\bar{C} + A\bar{B}C + \bar{A}\bar{B}C + ABC$. This expression should be mapped to see whether any simplification is possible. The mapping is:

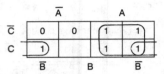

From the map, $F = A + \bar{B}C$. Once the minimal Boolean expression has been obtained it must be implemented using the appropriate gates. Three possibilities exist. The wanted circuit can be implemented

    (a) using AND, OR and NOT gates
    (b) using NAND gates only, or
    (c) using NOR gates only.

In both the TTL and the CMOS logic families NAND and NOR gates are cheaper, faster to operate and dissipate less power than AND or OR gates. It is also advantageous from both the manufacturing and maintenance points of view if a circuit uses just one type of gate throughout. It is therefore common practice to construct a circuit using either NAND or NOR gates alone.

The next step in deriving the wanted circuit should be to draw the logic diagram using AND, OR and NOT gates. The diagram can then be converted to a circuit that uses either NAND or NOR gates only by simply replacing each AND/OR/NOT gate by its NAND/NOR equivalent.

The final step is then to decide which integrated circuits will be used to construct the final circuit.

Suppose the Boolean expression to be implemented is $F = A + \bar{B}C$.

    (a) The logic diagram using AND/OR/NOT gates is shown in Fig. 8.24. One AND, one OR and one NOT gate are needed, necessitating the use of three integrated circuits.
    (b) The NAND equivalent of Fig. 8.24 is obtained by replacing each gate by its NAND gate equivalent. The circuit obtained is shown in Fig. 8.25(a). It might seem, at first sight, that six gates are now needed but two of these gates give successive

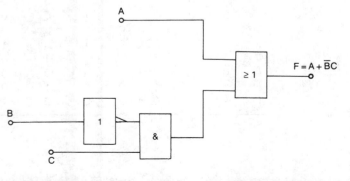

Fig. 8.24

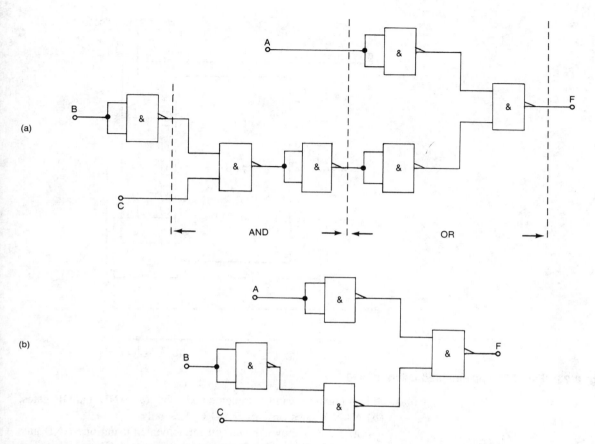

(a)

(b)

Fig. 8.25 NAND gate implementation of Fig. 8.24

inversions and are therefore redundant. Removing the redundant gates gives the final circuit of Fig. 8.25(b). This circuit contains four NAND gates and will need only one integrated circuit. Fig. 8.26 shows the circuit constructed with a 7400 quad 2-input NAND gate.

(c) The NOR gate version of Fig. 8.24 can be obtained similarly. Fig. 8.27(a) shows the first logic circuit diagram and Fig. 8.27(b) shows the final circuit after redundant gates have been removed. Again, four gates are needed and the circuit could be fabricated using the 7402 quad 2-input NOR gate.

Logic Circuits That Meet a Given Specification

1. A machine is to start working only when a start button is pushed by an operator (A), OR a start signal is received from a control point (B), AND a protective guard is in place (C) AND the piece to be machined is in position (D). The Boolean equation describing this action is:

$$F + (A + B)CD$$

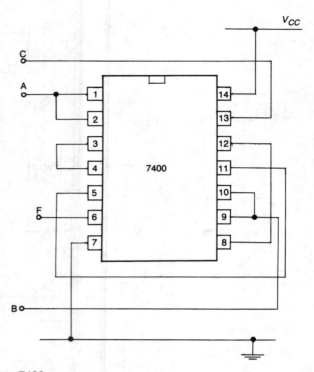

V_{CC}

Fig. 8.26 Fig. 8.25 implemented using a 7400

This equation can be implemented using (a) AND and OR gates, (b) NAND gates only or (c) OR gates only.

Fig. 8.28(a) shows the circuit implemented using one AND gate and one OR gate which requires the use of two ICs.

Referring to Fig. 8.23 and then replacing the OR gate with its NAND gate alternative, and from Fig. 8.23 also replacing the AND gate with two NAND gates, gives the circuit shown in Fig. 8.28(b). The circuit requires the use of five NAND gates and so could be implemented using two ICs.

The NOR gate version of the circuit is shown in Fig. 8.28(c). The OR gate has been replaced by two cascaded NOR gates and the AND gate has been replaced by three NOR gates.

2. A circuit is needed to operate the traffic light used in a toy car display in a shop window. Red, yellow and green LEDs are to be used as the lights which are to follow the UK sequence of red, red and yellow, green, yellow and then red again.

The truth table for the circuits is given by Table 8.19.

From Table 8.19

$$\text{red} = \bar{A}\bar{B} + A\bar{B} = \bar{B}$$
$$\text{yellow} = A\bar{B} + AB = A$$
$$\text{green} = \bar{A}B$$

Table 8.19

B	A	red	yellow	green
0	0	1	0	0
0	1	1	1	0
1	0	0	0	1
1	1	0	1	0

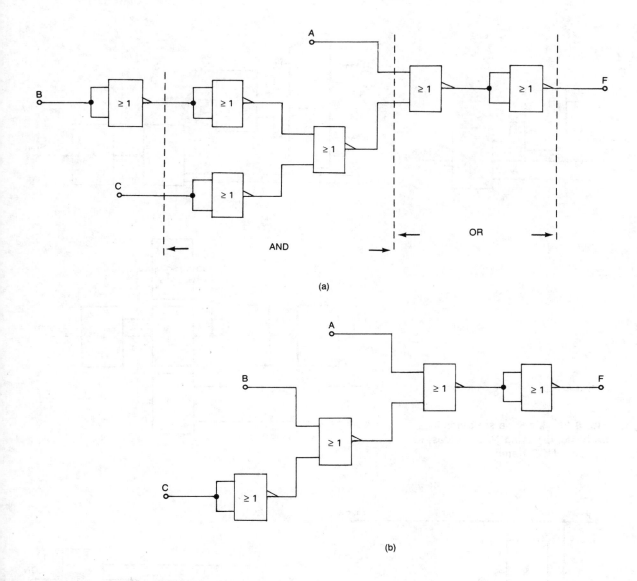

(a)

(b)

Fig. 8.27   NOR gate implementation of Fig. 8.24

The NAND gate implementation of the circuit is shown in Fig. 8.29.

3. A greenhouse heater is to turn ON automatically when (*a*) a switch (S) is in the ON position, AND (*b*) the outside temperature (B) has fallen to below a set value, AND (*c*) it is dark outside (C).

If B is high when the temperature is above the set value and C is high when it is daytime, then the Boolean equation that describes the operation of the circuit is $F = S\bar{B}\bar{C}$. The circuit is shown by Fig. 8.30.

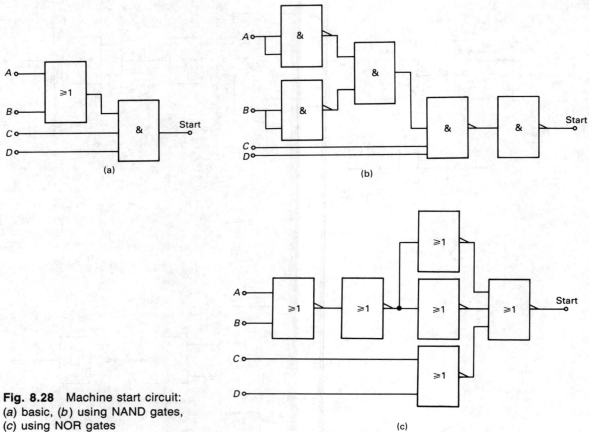

Fig. 8.28 Machine start circuit:
(a) basic, (b) using NAND gates,
(c) using NOR gates

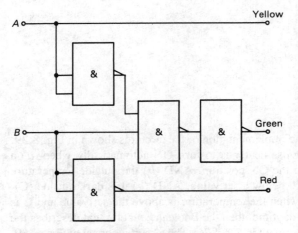

Fig. 8.29 Toy traffic light circuit

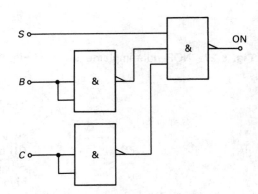

Fig. 8.30 Greenhouse heater circuit

Exercises

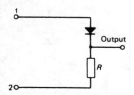

Fig. 8.31

Fig. 8.32

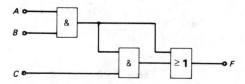

8.1 In the circuit given in Fig. 8.31 the load resistor R is 1000 Ω. The voltage applied between terminals 1 and 2 makes terminal 1 6 V positive with respect to terminal 2. When conducting, the diode has a voltage drop of 0.65 V. Calculate the output voltage.

8.2 When $A = B = 1$ and $C = 0$ find the output F of the circuit shown in Fig. 8.32.

8.3 The Boolean expression for a circuit is $F = ABCDE$. Use a truth table to show that the output of the circuit will be at logical 0 whenever $C = 0$ regardless of the values of the other inputs.

8.4 The Boolean expression for a circuit is $F = A + B + C + D + E$. Show that the output of the circuit will be at logical 1 whenever $C = 1$ regardless of the values of the other inputs.

8.5 The waveforms shown in Fig. 8.33 are applied to the circuit given in Fig. 8.34. Draw the output waveform F of the circuit.

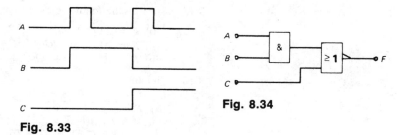

Fig. 8.33

Fig. 8.34

8.6 Write down the truth table of the circuit given in Fig. 8.32. Hence obtain the Boolean equation.

8.7 Write down the truth table of the circuit given in Fig. 8.34. Hence obtain the Boolean equation.

8.8 For the circuit given in Fig. 8.35 determine the logical value of B if $F = 1$ when $A = 1$.

8.9 Write down (a) the truth table and (b) the Boolean expression describing Fig. 8.35.

8.10 The Boolean expression for a logic circuit is $F = A.B.C.D$. Show how the function can be implemented using (a) one gate and (b) two gates.

8.11 The waveforms shown in Fig. 8.33 are applied to Fig. 8.32. Draw the output waveform.

8.12 Draw circuits to perform the logical functions (a) $F = (A + B + C)(D + E)$, (b) $F = A.B.C + D(B + C)$.

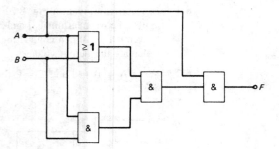

Fig. 8.35

Table 8.20

A	+5 V	+5 V	0 V	0 V
B	+5 V	0 V	+5 V	0 V
F	+5 V	+5 V	+5 V	0 V

8.13 The voltage table for a logic circuit is given in Table 8.20. What logical function is performed if (*a*) positive and (*b*) negative logic is employed?

8.14 The Boolean expression $AR + \bar{B}C$ is to be implemented. Draw the circuit using (*a*) NAND gates, (*b*) NOR gates only.

8.15 Simplify the following expressions using a Karnaugh map

    (i) $F = ABC + B\bar{C} + \bar{A}\bar{B}\bar{C}$
    (ii) $F = ABC + \bar{A}BC + A\bar{C} + B$
    (iii) $F = \bar{A} + BC + A\bar{B}\bar{C} + A\bar{B}C + B\bar{C} + \bar{A}B$

8.16 Simplify
    (*a*) $F = (A + B)(A + C)$
    (*b*) $F = \overline{A + \bar{B}} + B$
    (*c*) $F = C(\overline{A + B + C})$
    (*d*) $F = \overline{(A + B) + (B + C)}$

8.17 Show that
    (*a*) $(A + B)(A + C) = A + BC$
    (*b*) $A + \bar{A}B = A + B$
    (*c*) $\overline{(A + B)C} + \bar{A}(\overline{B + C}) = \bar{A}\bar{B} + \bar{C}$
    (*d*) $A\bar{B} + A\bar{C} = \bar{A} + BC$

8.18 Use a Karnaugh map to simplify
    (*a*) $F = ABC + \bar{A}BC + \bar{A}B\bar{C} + AB\bar{C}$
    (*b*) $F = ABC + \bar{A}\bar{B}\bar{C} + \bar{A}\bar{B}C + \bar{A}B\bar{C} + \bar{A}BC$
        $+ AB\bar{C} + A\bar{B}\bar{C} + A\bar{B}C$

9 Sequential Digital Circuits

A sequential logic circuit is one that is able to store one bit, or more, of data and whose output depends on both stored data and new input data.

The basic sequential logic circuit is the *flip-flop*. A flip-flop is a circuit which has two stable states: either it is SET, i.e. its Q output is at logical 1, or it is RESET, i.e. its Q output is at logical 0. There are four kinds of flip-flop in use; these are known as the $S-R$, the $J-K$, the D, and the T flip-flops. A flip-flop may be used on its own in a circuit, when it is used as a 1-bit store, or it may be used in conjunction with one or more other flip-flops to form a counter or a shift register.

Flip-flops

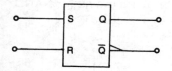

Fig. 9.1 *S—R* flip-flop

The $S-R$ Flip-flop

The $S-R$ flip-flop is a circuit that has two input terminals S and R, and two output terminals Q and $\bar{Q}$. In addition, an $S-R$ flip-flop may be clocked; this means that a rectangular pulse waveform, known as the *clock*, is applied to a third input terminal to determine the times at which the circuit changes state. The symbol for an unclocked $S-R$ flip-flop is shown by Fig. 9.1.

The truth table of an $S-R$ flip-flop is given by Table 9.1. In this table Q is the present state of the Q output terminal and Q^+ is the next state. At all times the $\bar{Q}$ terminal is complementary to the Q terminal, i.e. if $Q = 1$ then $\bar{Q} = 0$.

(a) When both the S and R input terminals are at the logical 0 level the flip-flop will remain in its present state, i.e. $Q^+ = Q$.

(b) When $S = 1$ and $R = 0$ the next state of the circuit will be $Q = 1$, $\bar{Q} = 0$, whatever the present state. The flip-flop is said to be SET.

(c) When $S = 0$ and $R = 1$, the next state of the circuit will be $Q = 0$, $\bar{Q} = 1$, whatever the present state. The circuit is said to be RESET.

Table 9.1

S	R	Q	Q⁺	
0	0	0	0	No change
0	0	1	1	
1	0	0	1	Set
1	0	1	1	
0	1	0	0	Reset
0	1	1	0	
1	1	0	X	Indeterminate
1	1	1	X	

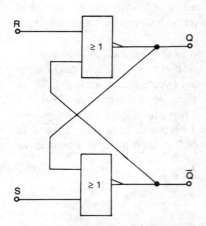

Fig. 9.2 NOR gate $S-R$ flip-flop

(*d*) When both the S and the R inputs are at 1 the flip-flop may, or may not, change states. The next state is *not* predictable and is said to be *indeterminate*. Such a condition cannot be tolerated in a practical system and if there is a possibility of the state $S = R = 1$ arising then the $S-R$ flip-flop should not be used.

The $S-R$ flip-flop can be obtained as an integrated circuit, e.g. TTL 7471 and CMOS 4043, but it can also be made using either NAND or NOR gates. Fig. 9.2 shows how two NOR gates can be interconnected to produce an $S-R$ flip-flop.

If two NAND gates are similarly connected, as in Fig. 9.3(*a*), the truth table has the 'no-change' and 'indeterminate' states interchanged. To obtain the true $S-R$ flip-flop logical operation two further NAND gates are needed, connected as shown by Fig. 9.3(*b*).

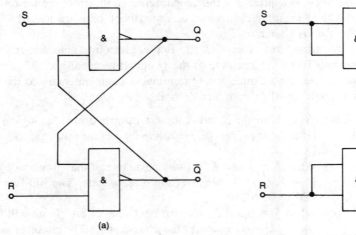

(a)

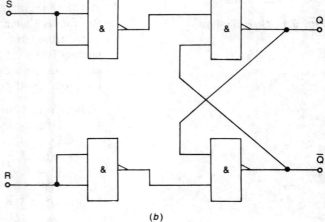

(b)

Fig. 9.3 NAND gate $S-R$ flip-flop

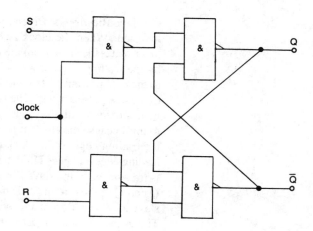

Fig. 9.4 Clocked NAND gate *S–R* flip-flop

Often it is desirable that the switching of the flip-flop occurs at defined instants in time that are specified by the *clock* (the clock is a rectangular pulse waveform). The NAND *S–R* flip-flop is easily modified to give clocked operation as shown by Fig. 9.4. Whenever the clock is at logical 0 the outputs of both of the input NAND gates must be at 1 regardless of the logical states of the *S* and *R* inputs. The circuit will then be unable to change state. Only when the clock input is at logical 1 will the *S* and *R* inputs control the operation of the circuit.

Table 9.2

J	K	Q	Q⁺	
0	0	0	0	No change
0	0	1	1	
1	0	0	1	Set
1	0	1	1	
0	1	0	0	Reset
0	1	1	0	
1	1	0	1	Toggles
1	1	1	0	

The *J–K* Flip-flop

The *J–K* flip-flop has a truth table which differs from that of the *S–R* flip-flop only in that the unwanted indeterminate state does not exist. The truth table of a *J–K* flip-flop is given by Table 9.2. It should be noted that when $S = R = 1$ the circuit *always* changes state or toggles.

A *J–K* flip-flop can be made by modifying the circuit of Fig. 9.4 but the resulting circuit is subject to unwanted *hazards*. Instead, one of the many integrated circuit *J–K* flip-flops is generally employed. Integrated *J–K* flip-flops are either *master–slave* or *edge-triggered* devices. The block diagram of a master–slave *J–K* flip-flop is shown in Fig. 9.5. The master flip-flop is directly driven by the clock but

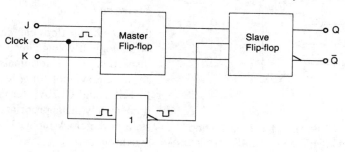

Fig. 9.5 Master–slave *J–K* flip-flop

the slave flip-flop is driven by the clock pulses after they have been inverted. When the clock is HIGH the inverted clock is LOW and this isolates the slave from the master. Both the flip-flops are triggered by the leading edge of a clock pulse. The master flip-flop responds to the data present at its J and K terminals as soon as the leading edge of a clock pulse arrives. The output of the master is then set, or reset, in accordance with the input data; the slave flip-flop does not respond to this because the clock at its clock input is at its trailing edge. When the trailing edge of the clock pulse occurs the inverted clock is at its leading edge and goes HIGH; at this point the data stored by the master is transferred to the slave flip-flop and appears at the output terminals Q and $\bar{Q}$ of the circuit. This means that the output state of a master–slave $J–K$ flip-flop changes at the trailing edge of the clock pulse. The J and K inputs must not change state while the clock remains HIGH.

Most IC $J–K$ flip-flops are edge-triggered. This means that they change state, as determined by the J and K inputs, *either* as the clock changes from 1 to 0 — a trailing-edge-triggered device — *or* as the clock changes from 0 to 1 — a leading-edge-triggered device. Leading-edge triggering is indicated by a wedge on the clock input of the flip-flop symbol, Fig. 9.6(*a*). Trailing-edge triggering is indicated by the addition of a small triangle, Fig. 9.6(*b*).

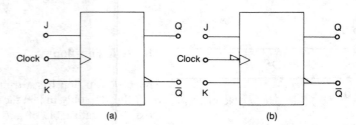

Fig. 9.6 Edge-triggered $J–K$ flip-flops: (*a*) leading edge, (*b*) trailing edge

(a)　　　　　(b)

The operation of the edge-triggered circuit is very quick and the J and K inputs do not need to be held constant while the clock is HIGH. For this reason edge-triggered $J–K$ flip-flops are used in preference to master–slave types whenever there is a chance that the J and/or K inputs may change during the duration of a clock pulse.

Preset and Clear

Some $J–K$ flip-flops can be set or reset (cleared) by the application of the logic 0 voltage level to its PRESET, or its CLEAR, terminal respectively. The terminals are said to be *active LOW*. Fig. 9.7 shows the symbol for a leading-edge-triggered $J–K$ flip-flop with PRESET and CLEAR terminals. The flip-flop is set, or is cleared, immediately the appropriate terminal is taken LOW regardless of the state of the clock input. The reset and clear actions are hence non-synchronous.

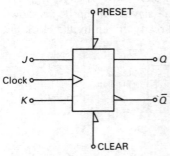

Fig. 9.7 Symbol for $J–K$ flip-flop with PRESET and CLEAR terminals

Table 9.3

D	Q	Q⁺
0	0	0
0	1	0
1	0	1
1	1	1

The *D* Flip-flop

The *D* flip-flop has one input terminal, a clock terminal and two output terminals, see Fig. 9.8(*a*). The truth table of a *D* flip-flop is given by Table 9.3. Clearly, the *Q* output always takes up the logical state of the *D* input. The *D* flip-flop is readily available in both the TTL and the CMOS logic families but it can also be obtained by modifying a *J*–*K* (or *S*–*R*) flip-flop in the manner shown by Fig. 9.8(*b*).

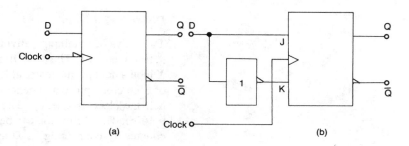

Fig. 9.8 *D* flip-flop

Table 9.4

T	Q	Q⁺
0	0	0
0	1	1
1	0	1
1	1	0

The *T* Flip-flop

The truth table of a *T* flip-flop is given by Table 9.4. It can be seen that every time there is a pulse at the *T* input the circuit toggles. The *T* flip-flop is not available as a separate integrated circuit: when one is wanted it is easily made by merely connecting together the *J* and *K* inputs of a *J*–*K* flip-flop. This is shown by Fig. 9.9.

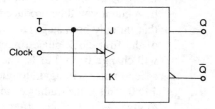

Fig. 9.9 *T* flip-flop

Counters

A counter is digital circuit that is able to count the number of pulses that are applied to its input terminals. It generates binary numbers in a specified count sequence. A counter goes through the specified sequence of numbers when it is triggered by an input pulse waveform and it advances from one number to the next only when a pulse arrives. The counter will go through the same sequence continuously as long as there is an input pulse waveform applied to its clock terminal. A counter consists of a number n of flip-flops connected together so that the output of one is applied to the input of the next. The number of states that a counter can take up is equal to 2^n and the highest number which can be stored is $2^n - 1$. A three-stage counter, for example, can have 2^3 or 8 different states and a maximum count of 7.

The number of different states is known as the *modulus* of the counter. A counter may be either synchronous (i.e. its operation is synchronized to the clock) or non-synchronous. The two most common counters are the decade counter and the binary counter. The decade counter has a count of 10; the binary counter has a count of 16.

Non-synchronous Counters

Divide-by-2 Counter

Two ways of obtaining a divide-by-two counter are shown by Figs. 9.10(a) and (b). The $J-K$ flip-flop circuit in (a) has both its J and K inputs held permanently at logical 1. The circuit toggles at the end of each clock pulse and generates an output pulse waveform at one-half the clock frequency. The circuit waveforms are given in Fig. 9.10(c). The D flip-flop can be converted to operate as a divide-by-2 counter by connecting its D and $\bar{Q}$ terminals together as shown in Fig. 9.10(b). The circuit's input and output waveforms are given in Fig. 9.10(d). At the first leading edge of the clock, when $Q = 1$ and $\bar{Q} = 0$ so that $D = 0$, the circuit switches to have $Q = 0$. Now $\bar{Q} = D = 1$ and on the leading edge of the second clock pulse the circuit switches again so that $Q = 1$, $\bar{Q} = 0$ and so on.

Divide-by-4 Counter

If two $J-K$, or two D, flip-flops are cascaded, Fig. 9.11, a divide-by-4 counter will be produced. Consider the $J-K$ circuit and suppose that initially both stages are reset ($Q_A = Q_B = 0$). The trailing edge of the first clock pulse (input pulse) will toggle flip-flop A and Q_A will change from 0 to 1. The other flip-flop will be unaffected. The second pulse will again toggle flip-flop A and make Q_A change from 1 to 0. This is the trailing edge of a pulse so that flip-flop B also toggles to give $Q_B = 1$. Thus, after two input pulses the state of the counter is $Q_A = 0$, $Q_B = 1$. A third input pulse will cause *FFA* to toggle again but the change from 0 to 1 in Q_A will have no effect on *FFB*. Now both flip-flops are set and $Q_A = Q_B = 1$.

A fourth input pulse toggles *FFA* and Q_A changes from 1 to 0; this change causes *FFB* to toggle also and so $Q_A = Q_B = 0$. The count sequence is now complete and any more input pulses will make the circuit count through the same sequence once again. The circuit waveforms are shown by Fig. 9.12.

Divide-by-8 Counter

A divide-by-8 counter requires three flip-flops connected in cascade and Figs. 9.13(a) and (b) show the $J-K$ and D flip-flop versions, respectively. This time the operation of the D flip-flop circuit will

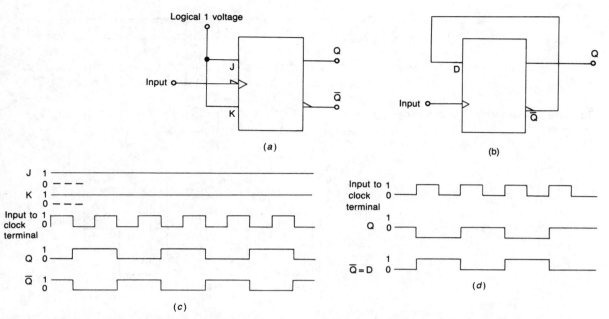

Fig. 9.10 Divide-by-2 circuits: (a) J–K flip-flop, (b) D flip-flop, (c) J–K waveforms and (d) D waveforms

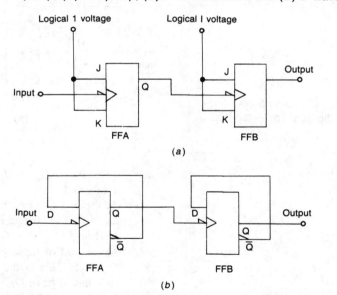

Fig. 9.11 Divide-by-4 counter: (a) J–K flip-flop, and (b) D flip-flop

be described. Suppose that all three flip-flops are initially reset so that $Q_A = Q_B = Q_C = 0$ and $D_A = D_B = D_C = 1$.

    (a) At the end of input pulse 1 *FFA* changes state to have $Q_A = 1$, $D_A = 0$. *FFB* and *FFC* are unaffected.

    (b) At the end of input pulse 2 *FFA* changes state to $Q_A = 0$, $D_A = 1$. The change in Q_A from 1 to 0 causes *FFB* to change

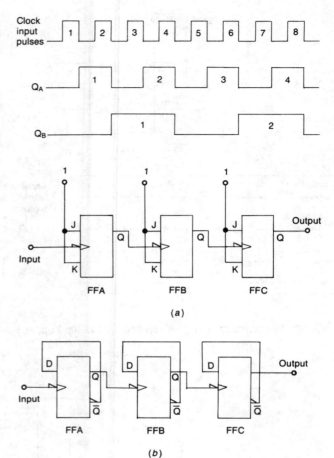

Fig. 9.12 Waveforms in a divide-by-4 counter

Fig. 9.13 Divide-by-8 counter: (*a*) *J–K* flip-flop, (*b*) *D* flip-flop

state. Now: $Q_A = 0$, $Q_B = 1$; $D_A = 1$, $D_B = 0$. *FFC* is unaffected.

(*c*) At the end of input pulse 3 *FFA* changes state to $Q_A = 1$, $D_A = 0$. The change in Q_A from 0 to 1 does not affect either *FFB* or *FFC*. Now $Q_A = Q_B = 1$, $D_A = D_B = 0$.

(*d*) At the end of input pulse 4 *FFA* changes state to $Q_A = 0$, $D_A = 1$. This is the trailing edge of a pulse so that *FFB* switches to have $Q_B = 0$, $D_B = 1$. The change in Q_B from 1 to 0 causes *FFC* to switch from RESET to SET and have $Q_C = 1$, $D_C = 0$. Now the counter state is $Q_A = Q_B = 0$, $Q_C = 1$.

The counter carries on with this sequence until the seventh input pulse is applied when $Q_A = Q_B = Q_C = 1$. An eighth input pulse will then reset all stages to give $Q_A = Q_B = Q_C = 0$. Thus the circuit has eight different states and a maximum count of 111 or decimal 7.

Synchronous Counters

Each of the counter circuits described so far has been a non-synchronous or *ripple* counter. Each flip-flop, other than *FFA*, cannot change state until the preceding flip-flop has changed state from 1 to 0. An input pulse appears to 'ripple' through the circuit and there is a cumulative delay in operation. Faster operation can be obtained if all the flip-flops can be made to change state simultaneously. A *synchronous* counter has all the flip-flop clock terminals connected together and to the input terminal so that all stages change state at the same instant.

Figs. 9.14(*a*) and (*b*) respectively show synchronous divide-by-4 and divide-by-8 counters.

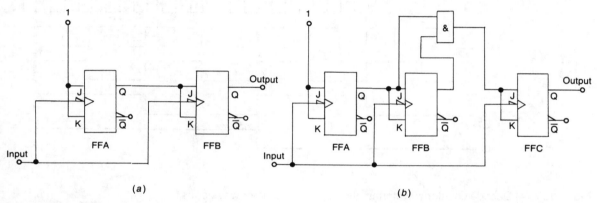

Fig. 9.14 Synchronous counters: (*a*) divide-by-4, (*b*) divide-by-8

The operation of the divide-by-8 counter is as follows. Assuming that all three flip-flops are initially reset the trailing edge of the first input pulse toggles *FFA* so that $Q_A = J_B = K_B = 1$. The next input pulse toggles both *FFA* and *FFB* to give $Q_A = J_B = K_B = 0$, $Q_B = 1$. Since $Q_A = 0$ the output of the AND gate is 0 and so are J_C and K_C. A third input pulse only makes *FFA* toggle. Now $Q_A = J_B = K_B = Q_B = 1$ and since both inputs to the AND gate are now 1 its output is also 1. Therefore, $J_C = K_C = 1$. The fourth input pulse makes all three stages toggle and produces the counter state $Q_A = Q_B = 0$, $Q_C = 1$. This means that $J_B = K_B = J_C = K_C = 0$ and so a fifth input pulse will toggle *FFA* only. The operation of the counter follows this sequence until an eighth input pulse is received; this will reset all the flip-flops and clear the circuit.

Reducing the Count

The maximum number of states which an *n*-stage counter can take up is equal to 2^n. The count can be reduced to less than 2^n in a number of different ways that are beyond the scope of this book.

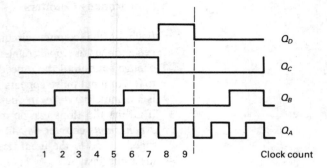

(a)

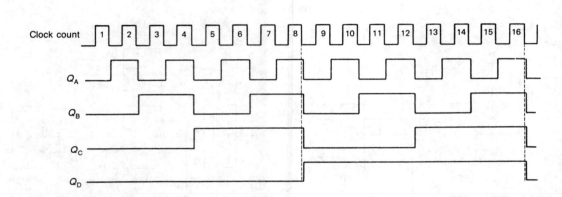

(b)

Fig. 9.15 (*a*) Decade counter waveforms and (*b*) binary counter waveforms

A decade counter counts a sequence of ten numbers from 0 through to 9. When the count is 9 the counter resets to return its count to 0. The waveforms at the four outputs of a decade counter are shown in Fig. 9.15(*a*). A binary counter counts a sequence of binary numbers ranging from 0 to 2^n where n is the number of stages in the counter. For a 4-bit counter $n = 4$ and the count is 0 to 15. Figure 9.15(*b*) shows the output waveforms of a 4-bit binary counter.

Example 9.1

(*a*) Determine the time taken for the decade counter to generate the count sequence if the clock period is 0.1 ms. (*b*) Determine the period and the frequency of each of the outputs Q_A, Q_B, Q_C and Q_D in Fig. 9.15(*b*) if the clock period is 0.1 ms.

Solution

(*a*) There are 10 pulses so time taken is 1 ms   (*Ans.*)
(*b*) Q_A: period = 0.2 ms, frequency = 5 kHz   (*Ans.*)
    Q_B : period = 0.4 ms, frequency = 2.5 kHz   (*Ans.*)
    Q_C: period = 0.8 ms, frequency = 1.25 kHz   (*Ans.*)
    Q_D: period = 1.6 ms, frequency = 625 Hz   (*Ans.*)

Table 9.5

Non-synchronous counters

7490	Decade	Up
7492	Modulus 12	Up
7493	Binary	Up

Synchronous counters

74160	Decade	Up	non-synchronous clear
74161	Binary	Up	non-synchronous clear
74162	Decade	Up	synchronous clear
74163	Binary	Up	synchronous clear
74190	Decade	Up/down	non-synchronous load
74191	Binary	Up/down	non-synchronous load
74192	Decade	Up/down	load/clear
74193	Binary	Up/down	load/clear

Integrated Circuit Counters

A number of IC counters are available in the various logic families. As an example, some of the more commonly employed TTL counters are listed in Table 9.5. The counters can be cascaded for applications that require more than four stages. An IC counter is provided with non-synchronous inputs that allow it to be preset to an initial value other than zero, or allow the counter to be reset back to zero at any time.

Shift Registers

A shift register is a digital circuit which can be used as a temporary store of data. It can be made using either $J-K$ or D flip-flops as shown by Figs. 9.16(a) and (b).

Data to be stored in a register is applied one bit at a time (or serially) to the data input terminal. It is loaded into the register by being shifted one place to the right at the end of each clock pulse. The number of bits of data that can be stored is equal to the number of flip-flops in a register. When the data is wanted it is moved out of the register one bit at a time by further right-shifting. This kind of shift register is of the *serial-in*, *serial-out* type (SISO). There are also three other ways in which a shift register may be operated. These are:

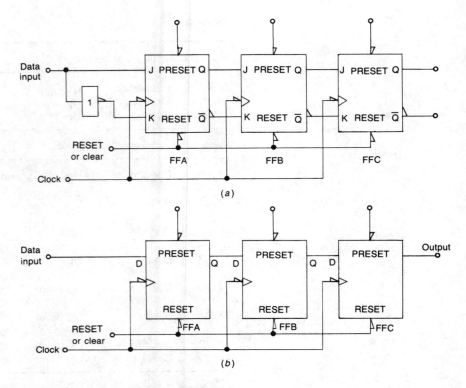

Fig. 9.16 (a) J–K flip-flop shift register, (b) D flip-flop shift register

(a) *Serial-in, parallel-out (SIPO).* Now data enters the register one bit at a time as for a SISO circuit but the data is taken from the register as an *n*-bit word with all bits simultaneously outputted. A SIPO register is often used to convert data from serial format to parallel format.

(b) *Parallel-in, parallel out (PIPO).* All the bits in the data are entered into the register simultaneously by setting, or clearing, each flip-flop, and they are also outputted simultaneously. Used as a temporary store.

(c) *Parallel-in, serial out (PISO).* With this kind of shift register data bits are simultaneously entered into the circuit but they are outputted one bit at a time as for a SISO circuit. Used to convert parallel data into serial form.

Both counters and shift registers are available in the TTL and the CMOS logic families. The counters are of either the synchronous or the non-synchronous kinds and may be 4-bit or 8-bit types. Any of them may be either a binary counter or a decade counter. A binary counter is the sort of counter previously described while a decade counter is a 4-bit counter that has had its maximum count reduced to 9 (0 to 9 is 10 different states) by resetting all the internal flip-flops when the count reaches 10; this takes the output of the counter to 0.

An up-down counter is able to count upwards from 0 to some maximum value, say 15, or downwards from 15 to 0. The direction of the count is determined by taking a count-up pin either high or low.

Exercises

9.1 Draw the circuit of a divide-by-8 counter using (a) synchronous and (b) non-synchronous techniques. Describe the operation of both circuits.

9.2 Explain, with the aid of truth tables, the difference between an $S-R$ and a $J-K$ flip-flop. Show how the $S-R$ flip-flop can be made using four NAND gates and how clocked operation can be arranged.

9.3 $J-K$ flip-flops are of either the master−slave or the edge-triggered type. Explain the basic differences between them.

9.4 Draw the circuit of a shift register that employs three D flip-flops. Briefly explain the action of the circuit when the input data to be stored is 101.

9.5 What logic level appears at the Q output of a flip-flop when it is (a) SET, (b) CLEARED, (c) RESET. A $J-K$ flip-flop has $Q = 1$. What is the $\bar{Q}$ level (i) initially, (ii) after the flip-flop has been toggled?

9.6 Draw the circuit of an $S-R$ flip-flop using the TTL 7400 NAND gate.

9.7 The binary signal 1010 is shifted into a 4-bit SISO shift register. What is the state of each of the four flip-flops after (a) 2 clock periods, (b) 4 clock periods and (c) 6 clock periods?

9.8 The binary signal 1101 is applied to a SIPO shift register. Draw the input and output waveforms.

9.9 A division of (a) 100, (b) 64 is required. Suggest how each could be obtained. What is the maximum count in each case?

9.10 (a) What is the main difference between a synchronous counter and a non-synchronous counter? (b) What is meant by the terms leading-edge triggered and trailing-edge triggered? Illustrate the answer with waveforms. (c) What is meant by the modulus of a counter? What is the modulus of a 4-bit binary counter?

Answers to Numerical Exercises

1.1 2, 3
1.2 29, 1, 29
1.3 2, 5
1.5 2, 8, 18
1.6 6.408×10^{-19} C
1.8 0.402 Ω-m
1.13 zero, decreased, positive on p-type
1.16 2.7 mA
1.17 0 mA
1.18 high, increased, assist
1.19 decreased

2.1 0.6 V
2.2 0.3 V
2.3 silicon, 120 V
2.4 0.27 Ω, 0.044 Ω
2.5 2.75 W
2.8 4.8
2.9 143 mA
2.10 40 pF
2.11 0.021 Ω
2.12 160 mA or 147 mA
2.14 silicon, 65 V
2.15 4 Ω, 0.03 Ω, power
2.16 5.6 V, 8 Ω

3.1 32 μA
3.2 3.1 mA
3.3 (c)
3.6 142
3.9 594 mV, 79.2
3.11 45 000
3.12 0.9972
3.13 0.9 mA
3.14 499, 499
3.15 52.08×10^3
3.16 25 μA, 25 mV
3.17 0.9975
3.18 121
3.19 65.7, 66.7
3.20 n-p-n
3.22 2.133 mA

3.24 33.3 mS, 154 mS, 200 mS
3.26 4.6 mA
3.27 n-p-n, 30 mS, 85 mS
3.28 59 kΩ, 10 kΩ
3.29 78 mS
3.31 (c)
3.32 33.3 mA
3.37 1.29 V, 0.93
3.39 80, 9600

4.3 4.5 mA
4.4 1.33 mS
4.5 40 kΩ
4.9 3.5 mS
4.10 4 kΩ
4.11 −1.1 V, very high
4.13 10 kΩ
4.14 3 mS
4.15 6 V
4.16 positive
4.17 10 MΩ
4.18 4.25 mS, 4 mS, 3.8 mS
4.19 n-channel dep. type MOSFET, 2.8 mS, 40 kΩ
4.21 1 V to 2.5 V depending on load
4.22 100 Ω
4.24 500 kΩ, n-channel dep. type MOSFET
4.25 1.05 mS, 500 kΩ

5.1 3600 Ω
5.2 240 Ω/□, 120 Ω/□, 60 Ω/□
5.3 1920 Ω, 3200 Ω
5.4 1250 Ω
5.6 200 mil²

6.1 5 mA
6.5 0.7, 0.6, −1 V
6.7 2 mA, 3 V
6.8 1.8 V
6.9 6.3 V, −1 V
6.10 5 V

6.11 11.5 μA
6.12 15 V, 2 V
6.13 6 kΩ, 1 mA
6.14 18 μA, 2.78 V
6.15 4.56 mA
6.16 600 Ω, 2400 Ω, 20 μA, 50.75 kΩ, 9.25 kΩ
6.17 25.6 mW, 11.4 mW
6.18 2.4 mA
6.19 0.69 V
6.21 4 mA, 18.8 V, 188
6.22 40 kΩ
6.23 2 V, 10 V, 12 V, 5.5 V
6.24 625 mV, 10.5 V, 3.7 mA, 266, 259
6.25 0.83 V, 33 mS
6.26 3 mS, 5.5, 6
6.27 100 Ω
6.28 3.9 W, 1.95 W

7.1 44 V, 8 V
7.2 230 V, 170 mA, 108 mA
7.3 120 V, 60 V
7.4 2.5%
7.6 99.06 V
7.7 16 Ω
7.8 yes
7.9 48 Ω minimum
7.10 490 Ω, 6.33 mA
7.11 105.6 mA

8.1 5.35 V
8.2 1
8.6 $F = AB$
8.7 $F = \bar{A}\bar{C} + \bar{B}\bar{C}$
8.8 1
8.9 $F = AB$
8.13 OR, AND
8.15 $AB + A\bar{C}$, $B + A\bar{C}$, 1
8.16 $A + BC$, $A\bar{B}$, 0, $\bar{A}\bar{B}\bar{C}$
8.18 B, 1.

Index

acceptor atom   10
a.c. resistance of a diode   17
a.c. load line   24, 131
aluminium   5, 9, 16
amplifier, audio-frequency   114
  bandwidth   123
  bias   119
  choice of configuration   115
  choice of operating point   116
  Class A operation   117
  Class B operation   118
  Class C operation   118
  current gain   129
  d.c. stabilization   122
  design of   140
  equivalent circuit
    h parameter   136
    mutual conductance   137
  gain using load line   127, 131
  integrated circuit   111
  multi-stage   142
analogue switch   89
AND gate   166
antimony   9, 16
array of transistors   68
atom 1
  acceptor   10
  aluminium   5
  boron   9
  donor   9
  germanium   4, 5, 6
  helium   3
  hydrogen   3
  impurity   9
  indium   9
  silicon   4, 5, 6
atomic number   3
avalanche effect   12, 27

bandwidth of amplifier   123
barrier potential   11
base transmission factor   42
bias circuit
  bipolar transistor   119
    collector-base   121
    fixed   120
    potential divider   121

FET   125
binary code   164
bipolar transistor   40
  action of   41
  array of   68
  base transmission factor   42
  characteristics   48
    input   49
    mutual   53
    output   51
    transfer   50
  collector leakage current   41, 53
  common-base circuit   43
  common-collector circuit   47
  common-emitter circuit   45
  construction of   63
  current gain
    h_{FB} 43
    h_{fb} 43
    h_{fc} 47
    h_{FE} 45
    h_{fe} 45
  cut-off frequency   55
  data sheet   55
  emitter injection ratio   41
  equivalent circuit   136
  fall-time   59
  frequency characteristic   55
  gain-bandwidth product   55
  h parameters   136
  input resistance   49
  integrated   102
  mutual conductance   54
  output resistance   51
  power gain   46
  rise-time   61
  saturation voltage   56
  selection of   66
  speed   59
  switch   56
  thermal runaway   55
  transition frequency   55
  turn-off time   60
  turn-on time   59
  voltage gain   46, 48
bit   121
boron   9

breakdown voltage of doide   12
bridge rectifier   150

capacitance of p-n junction   12, 29
capacitor, IC   107
capacitor-input filter   153
characteristics
  bipolar transistor   48
    input   49
    mutual   53
    output   51
    transfer   50
  diode   18
  FET   82
    drain   82
    mutual   84
  IGBT   94
  MESFET   95
  VMOSFET   92
charge carrier   7, 10
chip   101
choke-input filter   153
clamping   36
  circuit   36
Class A, B and C operation   117
CMOS   112, 164, 170, 173
combinational
  digital circuits   164
  logic   165
  logic circuits   176
    design of   178
    examples of   184
    use of NAND/NOR gates   181
combinations of gates   171
common-base connected transistor   43
common-collector connected
    transistor   47
common-emitter connected transistor   45
conductor   1, 8
counter   195
  IC   201
  non-synchronous   196
  synchronous   199
covalent bond   6
current
  diffusion   11
  drift   10

electron 12
hole 12
minority charge carrier 10
reverse saturation 12, 20, 41
cut-off frequency 55

D flip-flop 195
dark current 30, 62
data sheet
bipolar transistor 65
diode 30
d.c. load line 21, 127
d.c. resistance of diode 18
De Morgan's rules 179
depletion layer 11
depletion-type MOSFET 59
design
of audio amplifier 140
of combinational logic circuit 178
diffusion 10, 14, 16
digital circuit 112, 170
diode 15
a.c. resistance 17
breakdown voltage 21
capacitance of 29
characteristics 18
construction of 15
forward current 20
forward voltage drop 20
integrated circuit 107
junction 16
light-emitting 29
load line for 22
photo- 30
PIN 33
planar 16
power 27
reverse recovery time 25
reverse saturation current 20
Schottky 34
signal 26
switch as 24
varactor 29
Zener 27
DMOSFET 93
donor atom 9
doping 9

electronic gate 165
element 1
electron 3, 5
current 12
emitter follower 48, 116
emitter injection ratio 41
enhancement-type MOSFET 80
equivalent circuit
bipolar transistor 136
h parameter 136
mutual conductance 137
FET 140
exclusive-NOR gate 170
exclusive-OR gate 170
extrinsic semiconductor 9

fall-time of transistor 59
field-effect transistor 75
characteristics
drain 83, 92, 94, 95
mutual 84, 92
data sheets 95
depletion-type MOSFET 79
DMOSFET 93
enhancement-type MOSFET 80
equivalent circuit 140
frequency effects 86
handling of MOSFET 87
integrated circuit 103
junction 75
action as an amplifier 77
construction 78
I_{DSS} 79
parameters 78
pinch-off 77
input resistance 79, 81
mutual conductance 78
power 90
switch as 87
switching circuit 88
temperature effects 86
threshold voltage 81
filter, rectifier circuit 149
flip-flop 191
D 195
J−K 193
S−R 191
T 195
full-wave rectifier 149

gain-bandwidth product 55
gallium 9, 16
gate 165
AND 166
exclusive-NOR 170
exclusive-OR 170
NAND 168
NOR 169
NOT 165
OR 167
symbols 165
germanium 1, 4, 6

h parameters 136
half-wave rectifier 148
heat sink 56
helium 3
hole 7
current 12
hole-electron pair 7
hydrogen 37

IGBT 93
impurity atom 9
indium 5, 9, 16
insulator 1, 8
insulated gate bipolar transistor 93
ion 8
integrated circuit 10, 12

bipolar transistor 102
capacitor 107
digital 112, 170
diode 104
fabrication of 108
JFET 104
linear 110
MOSFET 103
resistor 105
intrinsic semiconductor 7

J−K flip-flop 193
junction
p-n 10, 11 12, 15
diode 16
FET 75

Karnaugh map 181

lateral p-n-p transistor 103
LED 29
lifetime of a charge carrier 7
light-emitting diode 29
line regulation 152
linear IC 110
load line 21, 127, 131
load regulation 152

majority charge carrier 10
MESFET 94
minority charge carrier 10
molecule 1
MOSFET 79, 80
multi-stage amplifier 142
mutual conductance 54, 78, 82, 84

NAND gate 168
neutron 3
non-synchronous counter 196
NOR gate 169
NOT gate 165
n-type semiconductor 9
nucleus 3, 5

op-amp 111
operating point of an amplifier 117
OR gate 167
output resistance
bipolar transistor 50
FET 79
over-voltage protection 162

peak inverse voltage 149, 151
peak reverse repetitive voltage 21
Periodic Table of the Elements 1, 5, 8, 9
phosphorous 9
photo-diode 30
photo-transistor 62
PIN diode 33
pinch-off 77
planar diode 16
planar transistor 63

p-n junction   10, 11, 12, 15
potential barrier   11
power
  diode   27
  gain of transistor   46
  MOSFET   90
  supplies   149
proton   3
p-type semiconductor   9

recombination   7
rectifier circuit
  bridge   150
  filter   151
    capacitor input   152
    choke input   153
  full-wave   149
  half-wave   148
  regulation   152
  voltage multiplier   153
  voltage regulator   154
    bipolar transistor   156, 158
    IC   159
    Zener diode   154
resistor in IC   105
reverse recover time   25
reverse saturation current   12, 20, 41
ripple voltage   149
risetime of a transistor   60

saturation voltage   56, 87
Schottky diode   34

semiconductor   1
  acceptor atom   10
  barrier potential   11
  capacitance of p-n junction   12, 29
  covalent bond   6
  depletion layer   11
  diffusion   11
  diode   15
  donor atom   9
  doping   9
  extrinsic   9
  hole   7
  hole-electron pair   7
  impurity   9
  intrinsic   5
  lifetime   7, 10
  n-type   9
  p-n junction   10, 11, 12, 15
  p-type   9
  potential barrier   11
  reverse saturation current   12
  thermal agitation   6
sequential logic circuits   191
  counters   195
  flip-flops   191
  shift registers   201
shift register   201
signal diode   26
silicon   1, 4, 6
$S-R$ flip-flop   191
stabilized power supplies   154
storage delay   59

swinging choke   153
switch
  bipolar transistor   56
  diode   24
  FET   87, 89
switching circuit   60, 88
synchronous counter   199

T flip-flop   195
thermal agitation   6
TMOSFET   93
TTL   112, 164, 170, 172, 201
turn-off time   60
turn-on time   59
two-state device   165

valence electron   5, 9
varactor diode   29
VMOSFET   90
voltage
  gain
    bipolar transistor   46, 48
    FET   134, 140
  reference diode   28
  regulator   154
    emitter follower   156
    IC   159
    transistor   158
    Zener diode   154

Zener diode   27
Zener effect   12, 27